▶ **Compromising the Ideals of Science**

DOI: 10.1057/9781137519429.0001

Other Palgrave Pivot titles by Raphael Sassower

DIGITAL EXPOSURE
Postmodern Postcapitalism

THE PRICE OF PUBLIC INTELLECTUALS

DOI: 10.1057/9781137519429.0001

palgrave▸pivot

Compromising the Ideals of Science

Raphael Sassower

University of Colorado, Colorado Springs, USA

DOI: 10.1057/9781137519429.0001

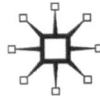

First published 2015 by
PALGRAVE MACMILLAN

Palgrave Macmillan in the UK is an imprint of Macmillan Publishers Limited, registered in England, company number 785998, of Houndmills, Basingstoke, Hampshire RG21 6XS.

Palgrave Macmillan in the US is a division of St Martin's Press LLC, 175 Fifth Avenue, New York, NY 10010.

Palgrave Macmillan is the global academic imprint of the above companies and has companies and representatives throughout the world.

Palgrave® and Macmillan® are registered trademarks in the United States, the United Kingdom, Europe and other countries.

ISBN: 978–1–137–51943–6 EPUB
ISBN: 978–1–137–51942–9 PDF
ISBN: 978–1–137–51941–2 Hardback

A catalogue record for this book is available from the British Library.

A catalog record for this book is available from the Library of Congress.

www.palgrave.com/pivot

DOI: 10.1057/9781137519429

▶ *Dedicated to the memory of Gary Orgel*

DOI: 10.1057/9781137519429.0001

Contents

DOI: 10.1057/9781137519429.0001

DOI: 10.1057/9781137519429.0001

Preface

Scientists used to be considered angelic by the public at least to the extent that they undertook their research for the love of God and human knowledge, rather than for any personal gain that may come about from the fruits of their inquiries. The acquisition of divine knowledge has been sanctified by a public that still worships the genius of scientists and the success of their endeavors over the past few centuries. Under ideal conditions of abundance (the fact that the 17th century Gentlemen of Science were independently wealthy) akin to the conditions of the Garden of Eden, we'd expect scientists to behave well and do only good. What happened to them along the way? Have they eaten from the forbidden fruit of the Tree of Knowledge and sinned in some biblical sense? Or was it the forbidden fruit of the trees of greed and fame that led them astray? Has their rank expanded too much to include some unsavory members, or have social demands on them become too onerous? Perhaps their aloof distance from the affairs of the state and their own institutional arrogance hastened their fall from grace, so to speak. There is some good and needed distance from the affairs of the state, as we shall see later, so that objectivity and detachment from political or financial influences is preserved; yet there is some detachment that is far too distant from the needs and concerns of one's fellow citizens. Though different, the two kinds of distance may be confused by both practicing scientists and their audiences.

In a 1988 PBS program titled "Do Scientists Cheat?" scientist Rustum Roy testified how he had to submit almost

DOI: 10.1057/9781137519429.0002

two proposals a week in order to support his laboratory. He also admitted that scientists regularly hype the promise of their research in order to get funded. A few years later, he was quoted in *Newsweek* describing scientists as having become "welfare queens in white coats." Roy clearly felt that many scientists were compromising the integrity of their research by placing it in the service of manna from heaven, or more realistically, federal or private funding. This book attempts in part to discern whether or not this admission reflects individual or institutional weakness.

In May 2013, John MacDougal, the President of the National Research Council of Canada, proclaimed publicly that "scientific discovery is not valuable unless it has commercial value." This is quite a far cry from the expectation of science as the venue through which we search nature's mysteries and the fundamental principles that govern it: we strive to conceptually understand an ordered universe and thereby predict its evolution. Is MacDougal's assessment a lamentation? Does the public indeed expect that much of scientists? Of course, the expectation of science to be perfect in some sense differs from the expectation that all scientists must likewise be perfect; some error and failure is expected of scientists, but on the whole, the expectation of science as a general practice and a calling remains at a much higher level.

The fall of scientists from grace differs from the biblical one, but, as I hope to illustrate here, it's a similar tale of woes. First, an (exaggerated) ideal view of the conditions of scientific research remains the backdrop against which scientists are judged today (Chapter 1). This view might be mistaken and overblown, but it anchors public perception and the decisions the state makes in light of it. Second, the socioeconomic conditions of scientific work have radically changed over the last three centuries, which warrants a reconsideration of public expectations of scientific research (Chapters 2–4). No longer are we funding research for the love of knowledge and to satiate human curiosity, but instead we have become beholden to national security concerns and the profitability of corporate interests. And finally, despite changing expectations, there is some hope for a renewed integrity of scientific work in the 21st century (Chapter 5), perhaps because of the realization that the expectations of scientists were too high to begin with, and that some of their compromises were and are still avoidable given the context of their practice.

I'm less concerned with the particular cases of fraud that periodically hit media outlets (even though some will be cited here as symptomatic of the community of scientists), and much more with the structural and

DOI: 10.1057/9781137519429.0002

institutional conditions under which individual scientists practice their trade. As becomes clear below, this line of argument is pursued with an emphasis on the institutional authority of science and its power to inspire young researchers as well as a public hungry for solutions to its common problems. Not all scientists are guilty of betraying their integrity just as not every citizen has clear-eyed and informed expectations of what the scientific enterprise should ideally look like. Yet rhetorical and discursive expressions in contemporary culture (see interesting examples on YouTube) make it abundantly clear that certain expectations remain intact: scientists should be angelic at least in the sense of maintaining personal integrity and fulfilling institutional codes of ethics, hovering, so to speak, above the fray of fear and greed, jealousy and rivalry. Just recall images of scientists in movies and plays, and in most cases we are introduced to eccentric, sometimes lovable, geniuses who may inadvertently fail, but their personal failures are excusable because of what their brilliance can teach us all.

My contention, then, is that public outrage voiced against the scientific community that has purportedly sold out is partially based on the false expectations of what science should ideally look like (given the changed political and economic landscape), and partially based on the unwillingness of the public to fund the scientific community without any strings attached. With a mistaken image of science and its practitioners, we are bound to be disappointed, frustrated, and at times just become outright angry.

There are numerous ways to approach the dependency of science on funding sources and the ways in which these sources – private, public, for-profit or non-profit – can and do exert their pressures to pursue specific agendas. Why else would a pharmaceutical company underwrite a biochemical laboratory if not in order to develop and market a new drug to treat a specific disease? With this in mind, it becomes obvious that the issue may not be so much *if* there is a financial relationship between scientists and their funding sources, but *under what conditions* do they force one kind of behavior as opposed to another. Funding scientific research as such isn't the issue; funding it for particular purposes may not be the issue as well as public and scientific interests may converge. Instead, what may be at stake is whether or not such funding necessarily or inevitably leads to the actual *corruption* of the ideals of science insofar as methods of inquiry are compromised, or that *expected* results are fabricated for the sake of continued funding. Here, too, I'm

DOI: 10.1057/9781137519429.0002

not contesting the cheating that goes on here and there, but rather the public relations feature of science in its misrepresentation of its ideals and practices. This means the kind of good-will marketing associated with all the benefits that come about because of scientific research. The notion of falling short of the changing societal expectations and the compromises scientists must make have to do as much with the changing social dynamics of the scientific community as with the changing socioeconomic conditions of democratic states, despite the immense authority the scientific community maintains.

Chapter 1 is devoted to a brief outline of the historical framing of scientific activities. The shift from modern *science* of the 17th century to the *scientific enterprise* of the late 20th century identifies a shift toward the importance of scientific funding and its rhetorical authority, a change in orientation toward the ways in which economic models (neoliberalism in particular) have infused meaning and strategies into the work of scientists. Understanding science as the amalgam of the activities of numerous scientists and scientific organizations, collectively the *scientific community*, means that apologists and critics alike are dealing with sociological issues and biases that motivate and direct the community instead of dealing with disembodied ideas (as feminists correctly point out) or rogue actors. I consider the cultural context of democratic capitalism as worthy of re-examination as well in this discussion – from market-capitalism to liberal democratic institutions – though I don't assume its universal application. Russia and China, for example, with their variant *state-capitalism* mode of governance produce more peer-reviewed research papers today than the US. So, what some may consider the necessary conditions for the proliferation of research (open channels of communication and entrepreneurship) need not be providing their sufficient conditions.

In Chapters 2–4, three main ways of changing expectations are examined. The very fact that scientific projects require funding isn't in dispute, nor that funding sources – government and private alike – expect accountability, if not an outright return on their investment. But what is of interest is whether funding does and should undermine the freedom and integrity of scientific research. Under what conditions is no harm done to those engaged in the process and their institutions, if in fact this is sometimes the case? Are the interests of the recipients of scientific data and technologies (from medical therapies and communication to digital gadgets and satellites) thoroughly anticipated and protected by the scientific community in

DOI: 10.1057/9781137519429.0002

terms of outlining potential dangers and pitfalls as well as by regulatory agencies? In other words, can science and the scientific enterprise function optimally or well enough despite some compromises by some of its practitioners? Obviously, funding alone isn't the problem, but rather the strings that are attached to such funding, the kinds that turn minor flaws into major questions of integrity. These strings simultaneously dictate which areas of research are chosen as well as the production and distribution of research results. Potential abuse is waiting at every turn of this convoluted path, but its impact isn't always similarly devastating. This is similar to a car engine that has a minor flaw or failure of one of its components without thereby bringing the car to a halt.

In reviewing some of the literature on the different contexts within which the scientific community operates, I was drawn to three main areas that characterize the chapters of this book. My own choices aren't as arbitrary as they may seem at first glance; they collect narratives and cases that can fit into these categories, thus illustrating the ways in which scientific practices compromise their own set of norms. For the sake of parsimony, I only deal here with Big Science as government-driven practice that may become nationalistic, and therefore eschews the universal nature of scientific knowledge; Big Money as either public or private agenda-setting that steers scientists in particular directions regardless of their own inclinations or the needs of the public; and Big Pharma as the clear expression of profit motives overshadowing any personal or communal scientific interests. The recent concerns with Big Data are only mentioned in passing in the concluding chapter insofar as the Internet and digital technologies challenge privacy and human dignity. In considering the changing expectations of scientists and the different compromises they have had to make so as to respond to them, I hope to also point to the different contexts where practical and moral quandaries are encountered in the age of neoliberal capitalism (which emphasizes the commercialization of every national activity, privatizing as much of the public domain as possible). Some contexts are more understandable, as in the case of government-directed research for national security; others are more pernicious (whether in the hands of government agencies or private entities) because of their secrecy and authoritative overreach. The consequences aren't simply harmful in particular situations but extend globally and to future generations.

Despite the inevitable failures and flaws we can expect to find in the scientific enterprise, I hope to illustrate in the concluding chapter that

DOI: 10.1057/9781137519429.0002

there is still hope for rescuing and reviving the ethos of science, that is, some of the norms that guide scientific inquiry. This will require going beyond the pathos of the ideological pronouncements of politicians and scientists and revisiting the economic conditions under which research and development may take place. Understood in our postmodern digital culture, we can envision a plurality of ways in which technoscientific work is undertaken, rather than the classical ones exemplified in the 20th century. Instead of only demanding a higher level of integrity from individual scientists, we should provide concrete proposals to institutionalize democratic reforms that can be supported by the public as a whole, reforms that account for different modes of practice and implementation, from start-up companies with few individuals to large webs of virtual connections without a centralized leadership.

DOI: 10.1057/9781137519429.0002

Acknowledgments

Special thanks go to my former teacher, Joseph Agassi, for steering me clear of some intellectual pitfalls, my brother-in-law, Sam Gill, for raising pertinent questions in relation to each chapter, my colleague, Jeff Scholes, for his concerns with the authority of science, my critic Carl Mitcham, for pushing me to clarify my arguments, Emma Longstaff of Polity Press, for encouraging me to think of this project, and the ever-dedicated and watchful Mia Tabib. Other helpful comments were provided by Margaret Foley, Shannon Hernandez, Seif Jensen, Eyal Kaplan, and Norman Roundy. Steve Fuller has been a constant inspiration to navigate the intellectual and publishing minefields of our day. As always, I'm grateful to my home institution because of its disinterestedness (in the Mertonian sense) in my research, allowing me to pursue my project without censorship.

DOI: 10.1057/9781137519429.0003

1

Fallen Angels: On the Compromises of Scientists

Abstract: *This chapter covers some of the ways in which science has been historically understood by the communities in which it flourished. It also explains the special ethos of science that was idealized in its modern formation and the ways in which the scientific community tried to adhere to its principles. Along the way, scientists fell short of the expectations set for them, and in time became more embroiled in the demands of the scientific enterprise. These demands brought about unseemly compromises that undermined the ideals of science.*

Sassower, Raphael, *Compromising the Ideals of Science*, Basingstoke: Palgrave Macmillan, 2015. DOI: 10.1057/9781137519429.0004.

Contemporary scientific endeavors cannot escape or be separated from legal, political, and economic spheres of influence (regimes of power in Michel Foucault's sense). Just think about the scientific enterprise as a process that includes at the very least the *production* of scientific ideas and then their application; the *distribution* of these scientific inventions as marketable devices or gadgets; and ultimately the *consumption* of these same objects by the public at large. Part of the production of scientific knowledge includes the education of students, and part of its distribution includes the challenges to religious and theological beliefs, and their consumption in contemporary culture. Therefore, each one of these three-staged events – the production, distribution, and consumption of science – deserves its own separate analysis. Such an analysis would include a deeper appreciation of the legal setting of each stage, from protection to regulation, from the secrecy of military research all the way to the use of medical devices and drugs with potential health benefits and risks. Likewise, such an analysis would include all the political maneuvering that influences state funding in relation to a specific constituency. The same can be said of the financial interests that surround anything with profitable potential.

As will become clear in this chapter, the romantic ideal of science for science's sake or the (Weberian-like) Ideal Type of science as a neutral activity the application of which remains mysterious and optimistically promises a better future has been debunked over the years. Do we still believe, as we may have years ago, that science is purely about unlocking nature's secrets? Are scientists indeed angels whose task is to connect the human realm with a transcendent realm full of wisdom? Some have argued that the role of scientists has traditionally been to bring us closer to understanding God's design of the universe, so that the presumed conflict of science and religion is a fiction of contemporary imagination. This line of questioning leads down a path that includes the religious dimensions of science or the realization that scientific research is at the service of a higher calling, but this path will not be examined here (Fuller 2007). Instead, this first chapter examines the alleged angelic or idealized view of scientists in order to appreciate the yardstick against which any present-day perception and behavior are measured. Whether or not we are justified in our condemnation of scientists as fallen angels, as falling short of what is expected of them, remains an open question. But it should be noted that this way of thinking rests broadly on the contestable premise that scientists were indeed like angels at some point

DOI: 10.1057/9781137519429.0004

in history or that they should aspire to be like angels in pursuing their vocation. No different from theologians who invoke the Garden of Eden with its peaceful abundance or political philosophers who postulate the "noble savage" in the State of Nature (Jean-Jacques Rousseau comes to mind in this context) in order to make recommendations for their own times, our view of science and scientists is idealized as well (no matter the reality that surrounds us).

The gentlemen of science

The shift from *science* to the *scientific community* in the 20th century and later to the *scientific enterprise* denotes more than a linguistic change (Greenberg 2001, 5–6). In fact, these subtle shifts in labeling express and announce a shift in the conscious self-description of what scientists do *qua* scientists. They belong to a community of scientists rather than being lone seekers of nature's truths with weird instruments and incomprehensible mathematical formulas. As community members, they are certified (education), indoctrinated (socialization), and follow the rules laid out by their professional gate-keepers (politics). They learn how to get funded (economics) and remain inside the permissible lines of conduct (law); they learn how to get published (reputation) and who not to antagonize (loyalty). Surely they also become proficient at their chosen disciplines, worthy of continued membership, but this proficiency is a *necessary* condition for belonging to the community; they still need to fulfill the *sufficient* conditions to ensure their continued membership. Given this description, are the ideals of science necessarily undermined? Science can survive the realities of the 20th century, as has recently been illustrated with the landing in 2014 of an unmanned spacecraft on a comet some 300 million miles away from earth. It can retain its lofty and inspiring role in our culture, accomplishing feats regularly postulated in sci-fi books and movies.

In order to understand the realities that face 20th-century scientists, it is important to retrace their roots or at least some of their predecessors. The "gentlemen of science" who set intimate relationships among themselves in 17th-century England believed in truth-telling as a matter of honor. For them, being considered gentlemen meant a deep level of trustworthiness that was earned as much as inherited. As the empirical sciences of that period developed, reports of sightings and experimental

DOI: 10.1057/9781137519429.0004

results had the credibility of those reporting them, and these gentlemen were wary of misrepresenting their findings as a matter of honor. Of course, some sloppiness crept into the scientific corpus of the day, but it was innocent as opposed to malicious (Shapin 1994, xxi, xxvii–xxxi). Science in the most general of terms meant an honorable pursuit, where mutual trust and gentlemanly conduct was expected.

My focus here on the gentlemen of science isn't meant to ignore the various original scientific thinkers from Galen (13–200) and Ptolemy (85–165) to Copernicus (1473–1543), Kepler (1571–1630), and Galileo (1564–1642), nor to render their practice in a nostalgic way so as to overlook the specific social preconditions of modern science in the UK. Instead, my intention is to highlight an institutional-like organization of various individuals and how their practices inform contemporary views and practices even when they are different in scope or intent. Significantly, these gentlemen of science valued *trust* as their common currency more than anything else, and an appreciation that the trust in knowledge was indeed dependent to a large extent on their trust in those providing the knowledge. This could be accomplished because these gentlemen's *virtue* was the warrant for truthfulness. Along these cultural lines of association and communication, Steve Shapin finds five reasons that justify the truthfulness of these gentlemen of science: competence, pragmatism, Christianity, disinterestedness, and self-control (Ibid. 67–86). It is therefore fascinating to note that the kind of gentlemanly conversation appropriate and expected in their social domain was copied into another, the newly emerging scientific discourse (Ibid. 122). In cases like these, we find that the surrounding culture influences the practices of the scientific community; they don't work in a social and moral vacuum; instead, they conform to well-established standards of behavior.

What is notable in this idealized conception of the production and distribution of science in the 17th and right into the 18th centuries, is that not only could one identify a set of criteria according to which trustworthiness can be ascertained – and penalized with ostracism when violated – but that one could even enumerate, as Shapin does, seven maxims for accepting testimony: testimony has to be plausible, multiple, consistent, immediate, it should come from knowledgeable or skilled sources, inspire justifiable confidence, and the individual sources must have earned their personal integrity and disinterestedness (Ibid. 212). This meant that one was required to fulfill these criteria in order for one's

DOI: 10.1057/9781137519429.0004

claims and reports to be considered scientifically credible. Perhaps we still uphold these criteria subconsciously, and are therefore mortified when members of the scientific community fall short of living up to them. This is where the public and its sub-communities share something in common: an expectation of conforming to ideals in the scientific discourse.

Not only was one supposed to behave in a gentlemanly manner in order to be taken seriously: there was a moral component – truth telling in everyday life – which was crucial for this new set of activities considered nowadays as the practices of natural scientists. As such, truth telling as expected in "civil conversation" was personally guaranteed, and therefore "knowledge was secured by trusting people with whom one was familiar, and familiarity could be used to gauge the truth of what they said" (Ibid. 410). Reports were delivered personally, and one's community was close by; it wasn't the anonymous Internet that acted as the communication channel, but intimate and friendly (and at times unfriendly) exchanges on a routine basis. These exchanges weren't supposed to bring about unanimity or a consensus of opinion, instead, they displayed a communicative and moral foundation that provided an intellectual safety net of some basic integrity.

Shapin reminds us that what distinguished the assent given to experts and to members within the scientific community, what he calls "members of scientific core-sets" or "modern techno-scientific knowledge-makers" was the fact that they know each other. This familiarity allows them "to hold knowledge as collective property *and* to focus doubt on bits of currently accepted knowledge," because it was "founded upon a degree and a quality of trust which are arguably unparalleled elsewhere in our culture" (Ibid. 417; italics in the original). There is an added moral pressure on the core group of experts because they feel accountable to their friends, to people they know and regularly interact with; they are not strangers whose approbation can be ignored. There are those who would argue Shapin's work is underpinned by a broader lesson, and counterintuitive finding, namely, that what we would today take to be self-evident common-sense values and organization of science were radically contravened by science at its early modern inception – so that the Mertonian "norms" of science that we may take for granted (discussed later) must also be sociologically contextualized.

Is this "premodern" British ideal a working heuristic with which to understand the critique and condemnation of contemporary practices?

DOI: 10.1057/9781137519429.0004

Can one claim that this has been true for the entire history of science with a few exceptions here and there? Or was this a unique condition where wealthy and privileged individuals could set their own agendas and fund them through family endowments, and therefore put their gentlemanly prestige ahead of personal interests or the interests of financial institutions? Was there something oddly unique about the socioeconomic conditions of the time that is fundamentally inapplicable either before or after that period?

To some extent, there was something unique about the conditions under which British science evolved from the 17th to the 19th century. We need only recall the experiences and work of Charles Darwin (1809–1882) who both subsidized his own research and who was fully aware of what reactions the results of his study would elicit. It's not simply that his views of evolution as random mutations and natural selection would undermine the Christian view of creation, but that his insistence on the randomness of the process challenges the religious and scientific quest for an ordered universe. Without going into the details of his research and their eventual reinterpretation in modern hands (for example, Gould 1977), it's worth emphasizing here that Darwin remains a complex figure, both as a British gentleman and as an innovative thinker who let the facts lead him to his radical conclusions. In this sense, he's both an angel and a fallen angel: rising above the fray to transcendent conclusions that would draw the admiration of some and the ridicule of others, and yet a tragic figure who believed he could reconcile theological beliefs with empirical data. The fact that he was able to finance his research himself and retain a level of scientific independence was as relevant in his time as it was for his predecessors, insulating him from the pressures of funding bodies or individuals. And as such, he provides another symbolic exemplar of what an ideal scientists should be like.

20th-Century ideals of science: from Weber to Merton

Max Weber, at the beginning of the 20th century, recast the notion of science for science's sake as "science as vocation." Taking into account the greater importance of institutions of higher learning and the fact that academics working in the natural sciences find themselves as part of universities (the German ones looking more and more like their American counterparts), Weber suggested that modern science is part

DOI: 10.1057/9781137519429.0004

of the "process of intellectualization" (1946/1922, 138). This process of intellectual "rationalization" is negative insofar as it is a never-ending process that destines its practitioners' work to always "be outdated," and does not "indicate an increased and general knowledge of the conditions under which one lives" (Ibid. 138–139). In addition to the specific impossible conditions under which scientists work, there is also an added sense that through their work – measuring and calculating – "the world is disenchanted," devoid of the meanings given to it by philosophy and theology, losing the mysterious features it should embody. So, what then is the "value" of science (Ibid. 140)?

Instead of mentioning some ideals of science as understood in the 17th century in the UK, Weber takes us all the way back to Plato and his story of the cave in the *Republic*. There, he reminds us, men were chained and could only see the shadows on the cave's wall rather than what really was happening behind them, outside the cave. The brave man who leaves the cave and sees the sunlight, who realizes that the world is larger than his cave, and that the images on the wall are not the real objects of nature, is, for Weber, "the philosopher"; the sun, in this retelling, is the "truth of science, which alone seizes not upon illusions and shadows but upon the true being" (Ibid.). Then Weber asks sarcastically or sadly, and of course rhetorically: "Well, who today views science in such a manner?" He answers: "Today youth feels rather the reverse: the intellectual constructions of science constitute an unreal realm of artificial abstractions, which with their bony hands seek to grasp the blood-and-the-sap of true life without ever catching up with it" (Ibid. 140–141).

The disenchantment of previous generations has come to its fullest expression in the workings of scientists who are now, under conditions of rationalization and intellectualization, detached from the reality of their lives. Just like Plato's "cavemen," they are lost once again in their scientific caves, their mathematical abstractions that bear no direct relations to the meaning of life which they ought to explain. Scientists' own claims about the worthiness of their activities are understood by Weber as mere presuppositions they themselves cannot defend, as they have given up on "living in union with the divine" or searching for "ultimate meaning" (Ibid. 142–143). If science doesn't ask anymore the difficult but fundamental questions of life, why even bother with any of its other answers?

With his critique of the value-neutrality claimed by scientific research and with his lament over the process of rationalization and abstraction

DOI: 10.1057/9781137519429.0004

that has made our world disenchanted, Weber provides yet another idealized view of what the scientific community ought to look like or be working toward. Just like the British gentlemen of science whose Christianity and class affiliation obliged them to behave in certain ways, so are 20th-century scientists supposed to seek the truth about life's meaning and retain their integrity (Ibid. 156). Spanning some three millennia, the ideal of science for science's sake, the ideal behavior of the seekers of truth, remains unshaken even when its impending loss is mourned.

This elevated view of the role of scientists filtered into literary circles in Europe by the end of the 19th century. We recall with fondness the internal turmoil of the country doctor who feels compelled to condemn the baths of his village as unsafe despite the economic benefits they are supposed to bring in Henrik Ibsen's play *An Enemy of the People.* Here stands Dr Stockman against the "people," arguing that "The most insidious enemy of truth and freedom among us is the solid majority" (1970/1882, 191). The lone man of truth, the country doctor as a man of science, facing the empirical facts of the situation, is set against the backdrop of ignorance and popular appeal, the "herd mentality" so eloquently described by Nietzsche around the same period.

Ibsen catches the sentiment of his day when casting his play in terms of truth versus lies, the single scientist who has integrity (Plato's courageous man who leaves the cave) versus the majority of the town (still chained in the cave) willing to overlook the truth for the sake of profit. Along the progress of the Industrial Revolution, certain problems became apparent, but only to those who were willing to address them. The problems of population migration and urbanization, environmental pollution and human suffering could escape those whose sole motive revolved around financial benefits, as Karl Marx loudly argued. But what about scientists who participated in the practical application of their research? Should they maintain their neutrality about the consequences of their rationalization and intellectualization? Or should they speak up like Dr Stockman and ruin their family's fortunes? Aren't they supposed to behave with greater integrity than their fellow capitalists?

Unlike Shapin's focus on the personalities that populate science and Weber's notion of the vocation of science that demands personal integrity and religious piety of scientists who are academics, we find, by the mid- and late 20th century, Robert Merton's description (which is also a prescription) of the ethos of science. As far as he is concerned – and this has become an article of faith (if not practice) ever since its formulation – scientific activity displays four major characteristics that define it as much as guide those engaged in it. These four features distinguish science from all other human activities. When thinking through these as a sociologist, Merton only partially follows his fellow sociologist, Weber. He does acknowledge the institutional trappings of science as well as those who participate in it – hence the shift from *science* to the *scientific community* – but the philosopher in him sets in place a classification that is more logical and methodological than Weber's nostalgia for the theological underpinnings of any vocation. Merton isn't concerned with the meaning of life or personal redemption, but rather with an institutional and professional adherence to a set of rules to ensure the self-policing of science unlike any other human endeavor.

The first feature of scientific inquiry is labeled "Universalism," which is supposed to capture four other terms associated with the workings of science, namely, that it is "impersonal, dispassionate, international, [and] democratic" (1973/1942, 270–273). With this phrasing, Merton cleverly pulls the psycho-social dimension out of science, suggesting that it ought to be impersonal and dispassionate. But, Weber could have asked him, how would you expect science to be one's vocation if the scientist isn't fully engaged and committed to the work? Perhaps a better question would be: what level of passionate engagement is warranted? It makes sense that science ought to be international and democratic, if by this Merton means that it knows no national boundaries or one's ethnic or religious background. In this, though, he does undermine the specificity of Shapin's British gentlemen of science, for they were anything but democratic or international. In fact, what made their work so successful by default was their parochialism. With the passage of time, better communication and traveling methods have enabled a more cosmopolitan approach to science. Moreover, during the 20th century, whenever the designation of science was "Jewish science" (as with the Nazis) or "bourgeois science" (as with the Communists), it was considered boorish and out of place. Merton's first feature therefore has been vindicated if not fully followed.

The second is "Communism" which he defines in three other related terms: "common ownership of property, collaboration, [and] standing on the shoulders of others" (Ibid. 273–275). To be perfectly clear, the communism or communalism advocated by Merton depends to a large extent on the ideals of communism (as explained by Marx), but isn't party oriented or beholden to the failed Soviet experiment. Instead, what

DOI: 10.1057/9781137519429.0004

this feature highlights is the fundamental commitment to sharing one's knowledge and findings with anyone else without any barriers (financial or other). Given the enormity of the tasks undertaken by scientists, and the large amount of data to be examined, measured, and calculated, it stands to reason that collaboration within and among laboratories should be common. The free flow of information would enhance the progress of science, allowing for greater numbers of testing of the same hypotheses (for potential confirmation or falsification) or the reformulation of auxiliary hypotheses when needed.

The classic Newtonian notion of "standing on the shoulders of giants" (Merton 1965) has been proven to be useful on many levels. To begin with, as students of our elders, we learn from them and therefore can (potentially but not necessarily) see farther than they were able to at their time. Secondly, this cumulative way of thinking about the history of science recognizes the contributions of predecessors and contemporaries alike and suggests that we are all beneficiaries of the knowledge of others. And lastly, as beneficiaries of scientific knowledge, we should consider ourselves as fellow travelers and contributors to the mass of accumulated knowledge. As such, we should freely dispense with all formalities and joyfully participate in unraveling the secrets of nature (or God for some who see the two as one and the same, as in Spinoza's sense).

The third feature is "Disinterestedness" which can be understood as an emphasis of an aspect already covered in the first two (Ibid. 275–277). How different is it from being "impersonal" or "dispassionate?" Perhaps what's at stake here is the practice, more fully elaborated on by Karl Popper with his notion of "conjectures and refutations" (1962), which encourages scientists to subject their own hypotheses to the scrutiny of others (rather than shield them from criticism). Would an ongoing process of critical inquiry reduce any potential for fraud? Would routine exposure to testing minimize (if not completely eliminate) fabrication? As far as Merton is concerned, "The virtual absence of fraud in the annals of science, which appears exceptional when compared with the record of other spheres of activity, has at times been attributed to the personal qualities of scientists." But what does this mean? Does it mean, for example, that scientists are "recruited from the ranks of those who exhibit an unusual degree of moral integrity?" This, of course, cannot be the case, because "there is, in fact, no satisfactory evidence that such is the case." And here is the Mertonian shift from a psychological or anecdotal to a sociological explanation: "a more plausible explanation may be found

DOI: 10.1057/9781137519429.0004

in certain distinctive characteristics of science itself. Involving as it does the verifiability of results, scientific research is under the exacting scrutiny of fellow experts ... the activities of scientists are subject to rigorous policing, to a degree perhaps unparalleled in any other field of activity. The demand for disinterestedness has a firm basis in the public and testable character of science and this circumstance, it may be supposed, has contributed to the integrity of men of science" (Ibid. 276).

Merton's "integrity of men of science" is quite different from Weber's pleading for vocational absorption, for it is institutionalized and publicized; it is a matter of public record that makes science both transparent and self-policing. So, instead of "men of science" bringing integrity to science, it is science that makes them so. This reversal of cause and effect is what makes science so unique and, as Merton believes, displays a "virtual absence of fraud." This, as we shall see in later chapters, is exactly what is being contested. But Merton's idealized summary and his deep faith in scientific integrity is what provides the ideological backdrop against which the very question of the failure of scientists to adhere to their own professional principles comes to light. We have encountered some comments to this effect in the Preface, and we'll see more of this sentiment in the later chapters.

The fourth feature is "Organized Skepticism," which for Merton means "a methodological and institutional mandate, temporary suspension of judgment, [and] detached scrutiny" (Ibid. 277–278). This last feature amplifies the notion of "disinterestedness" and switches the emphasis from the scientist who produces reports of natural phenomena to the community of scientists who are the recipients of such reports. Instead of welcoming fellow workers to collaborate and share in the excitement of the potential of these reports, they encourage an expectation for the scientific community to display an active organized skepticism. No matter the source, no matter the prestige of the source (in contradistinction to the mindset of the 17th-century gentlemen of science), they are supposed to be skeptical, doubtful that what is presented to them has the validity claimed by the scientist–author. Their doubt or skepticism, once again, isn't personal in nature, but "organized" institutionally. This means that journal referees judge blindly, without prejudice; it also means that the burden of proof remains fully on the source of the report, the scientist who makes a claim publicly. The authority of individual scientists or their laboratories (or institutions) is never accepted on faith alone; it must be earned after a critical and grueling process.

DOI: 10.1057/9781137519429.0004

Regardless of the difficulties associated with any of these four features (and by extension with all of them together), each remains the idealized characteristic that defines scientific inquiry. Communal through and through, this picture of science is still far removed from what we perceive to be the *scientific enterprise* which rears its overwhelming head by the late 20th century. Weber worried as early as the 1920s about the changing ethos of the university, becoming more beholden to the "practical" affairs and the financial dependency of its members (scientists included). He had no idea what was still to come. Merton, who lived through the post-World War II American landscape, should have been even more alarmed than Weber. Should we read his scientific ethos as a means to remind practitioners of the ideals of science? According to Joseph Agassi, the scientific community still adheres to an image of science that has rationality – "perfect proof" – as its basic foundation (2003, 153). Following Popper's view of science as a "social phenomenon," Agassi's main concern is to foster the autonomy of scientists. As long as the idealization of science enhances critical rationalism and isn't excessive, it is "legitimate" (Ibid. 161–162). The quest for knowledge, likewise, is legitimate so long as it doesn't require "certitude," which is as impossible to achieve as uniformity or unanimity among scientists (Ibid. 159). The ideals of science have been redefined in the 20th century, but despite some revisions here and there, they remain intact in the public's mind.

Contemporary views of the ideals of science: from Kuhn to Ziman

By contrast to both Weber and Merton, Thomas Kuhn paints a different picture of the *scientific community*, one beholden to leaders as teachers and to mentors as laboratory chiefs, and one in which paradigm indoctrination is the rule rather than the exception, as most of what happens is "normal science" rather than "revolutionary science." In his words: "Men whose research is based on shared paradigms are committed to the same rules and standards for scientific practice. That commitment and the apparent consensus it produces are prerequisites for normal science, i.e., for the genesis and continuation of a particular research tradition" (1970/1962, 11). What are the implications of this picture in contrast to earlier ideals? Perhaps descriptively more accurate and nuanced, Kuhn leads the scientific community down the path of conformity

DOI: 10.1057/9781137519429.0004

(through the consensus obtained in laboratories and within disciplinary boundaries), a path much less heroic or vocationally inspiring than the one advocated by the gentlemen of science, all the way to Weber's and Merton's formulations. Has Kuhn given up on their ideals and settled for "tradition" and "shared paradigms?" According to Steve Fuller, it's the climate of the Cold War that influenced his thought so much that any espousal of democratic ideals (Popper's and Agassi's) seems foolishly romantic or outright dangerous. As Fuller explains in wonderful clarity: "With the defeat of Popper (and his follower), the normative structure of science drastically changed. Whereas actual scientific communities existed for Popper only as more or less corrupt versions of the scientific ideal, for Kuhn the scientific ideal is whatever has historically emerged as the dominant scientific communities" (2004, 4).

So, the notion of paradigm has become a convenient means – with its honorific connotations – by which Kuhn justifies (rather critically assesses) an ever-increasing collection of complex or entropic data across national borders, keeping it together as a pragmatically appropriate arm of government goals and power relations. This means that under the conditions of the Cold War or perhaps as an idealized model, Kuhn's scientific community functions within the guidelines of the political agenda instead of remaining outside of it, free to pursue a line of research in whatever direction the paradigm leads it. Autonomy under such view means something completely different from what has been argued about for two centuries. It means autonomy within the confines of the state, perhaps within smaller and smaller sub-groups in a laboratory (Ibid. 50–51).

Writing at the dawn of the 21st century, the British John Ziman offers an updated version of Merton's four features and Weber's notion of vocation as the ideals of scientific research. He adds "Originalism" as an important feature of the "Republic of Learning," continuing the already accepted shift from *science* to the *scientific community* (here a republic). For Ziman, academic scientists are socialized to be "self-reliant and independent"; they are supposed to come up with novel answers to questions they themselves choose to research (unlike Kuhn's confinement to the boundaries of the state). Their originality is twofold: "At a mundane level, every scientific paper must contribute *something* new to the archive." At the level of "notable achievement" where the stakes are higher than just another refereed publication, originality has to be *"proved"* (2000, 40–41; italics in the original). One's ability to prove originality or the priority

DOI: 10.1057/9781137519429.0004

of discovery – played out in the urging of Darwin to publish his results before Alfred Russell Wallace – has a long and fascinating history. For science to claim a special status among all human activities, it must incorporate a strong sense of novelty, a progressive movement toward finding new and original answers and solutions to nature's unsolved paradoxes and anomalies – its secrets. And the way to make these answers public is through publication in both professional and popular outlets.

It is here, though, that Ziman reverts from the methodological to the sociological aspects that guide and drive the scientific community, especially when rivalry and competitiveness come into play: "But such conflicts do not prove that scientists are peculiarly jealous and contentious as *individuals*. They indicate, rather, that science is a powerful and intensely normative *institution*, where people rightly feel that proper behavior and outstanding achievement should be fairly recognized and rewarded" (Ibid. 42; italics in the original). Though describing the transformation of the scientific enterprise into its "post-modern" culture where pluralism (rather than reductionism) is the catchword and conceptual framework, Ziman remains beholden to the "normative" elements that make science an "institution" worthy of emulation and admiration (Popperian or Kuhnian?). In other words, instead of reducing all phenomena and discoveries into one set of causes or basic elements, Ziman endorses the pluralism of postmodern thought and practice, where multiple causes bring about multiple effects, and where there may not be found any singularity of an explanatory (and predictive) model. Justice and fairness are as important to one's scientific contributions as one's brilliance and authorship, perhaps even more so, as Ziman expects truthfulness and "proper behavior," much like the gentlemen of science of yesteryear. Even when the cases of the likes of Madam Currie come to mind as someone who was robbed of recognition during her lifetime (and for a while after her death, too), we'd hope with Ziman that her discoveries never lost their empirical significance for the promotion and development of science. As we shall see at the closing of the book, a pluralistic postmodern mindset even among the community of technoscience is becoming more apparent, without the downfalls of its relativistic features.

As we move to the 21st century, remnants of the idealized pictures of science are still with us. For example, Sheldon Rampton and John Stauber argue that "historically, science has often allied itself with the political philosophy of democracy" (2001, 33–34). This means that at

DOI: 10.1057/9781137519429.0004

times, science was in the service of the powers to be and at other times, it propounded the ideals of democratic and communal association of knowledge seekers that should inform society as a whole. But taking on this posture has meant that "science had ceased to be merely a methodology and had become an ideology as well" (Ibid. 37). There are some who contest this view of the democracy of science, because the search for the truth isn't accomplished by majority vote (Culshaw 2007, 18). On another level, one could argue that science has always had an ideology, especially as science was set against the backdrop of a powerful Catholic Church that demanded subservience to its dogmatic doctrine. Yet this scientific ideology, whether applied internally to its community or espoused as a model for society as a whole, is still dependent on the practices of actual working scientists to either confirm or demystify it, as Agassi reminds us. Besides, one can also argue that regardless of any ideology, there is something unique about human curiosity and its motivation to inquire and speculate, probe and experiment, under any and all political and economic conditions. Whether understood as an instinct that is part of human nature or an attribute rewarded by evolutionary forces, it's reasonable to conclude that scientific inquiry will persist no matter what happens to be its social or political organization.

This means that generalizing about the *scientific community* is dangerous and misleading. For some, what remains front and center when the activities of scientists are examined is the following fact: "The actual thought processes of scientists are richer, more complex, and less machine-like in their inevitability than the standard model suggests. Science is a human endeavor, and real-world scientists approach their work with a combination of imagination, creativity, speculation, prior knowledge, library research, perseverance, and, in some cases, blind luck – the same combination of intellectual resources, in short, that scientists and nonscientists alike use in trying to solve problems" (Rampton and Stauber, 2001, 197). Even when exceptional individuals are guided by the scientific ethos, these scientists operate just like nonscientists, with all the luck and success on the one hand, and the potential for failure and (unintentional) fraud or blunder on the other (Livio 2013).

Yet on many levels scientists still aren't like other professionals, because they are expected "By the traditions of their calling, and in popular understanding... to possess a selfless dedication to truthfulness, the growth of knowledge, and the public interest" (Greenberg 2007, 2). Perhaps what they face, as Daniel Greenberg emphasizes, is "the faith

DOI: 10.1057/9781137519429.0004

in scientific saintliness," regardless of whether or not it's historically justified. This is the kind of "saintliness" we have mentioned in passing in regard to viewing scientists like "angels" who might be considered "fallen" when they fail at their calling. Practically speaking, he continues, "Honest science, unsullied by commercial lures and pressures or politics, is clearly invaluable public good, whether for assessing the safety of drugs or the menace of global climate change" (Ibid. 4). The question, then, is whether or not we (and our scientists) are ready to shed the idealized images of science; there is a practical reason to perpetuate the mythology and elevate it to ideology – it's good for public policy and welfare (if not public relations, as Agassi argues). And when it gets to the Weberian sense of scientific integrity, Greenberg, like many others writing about the future of the *scientific enterprise* of the 21st century, still upholds the view that "Transparency, openness, and disclosure are the most frequently prescribed measures for bolstering scientific integrity" (Ibid. 102). Perhaps Weber was right after all to cling to personal integrity as the foundation for all the lofty features outlined by Merton.

We can even cite the heralded cases of Albert Einstein from the earlier part of the 20th century all the way to that of Stephen Hawking later in that century. Their genius and public persona have had much to do with how we have come to think about them if we ourselves aren't working scientists. They exemplify the angelic image we have of genius and selflessness, and their press coverage has always been adoring. Einstein's iconic image has been canonized in posters with his uncombed and unruly white hairdo, or with his tongue out, and even in a movie with his fellow geniuses at Princeton University's Institute for Advanced Studies ("I.Q." 1994). Similarly, the iconic image of Hawking in his wheelchair with a mind-activated computerized voice has become even more commonplace after the recent movie "The Theory of Everything" (2014). There are other, more tortured images of great scientists, from the American mathematician John Nash ("A Beautiful Mind" 2001) to Alan Turing, the British mathematician, logician, cryptanalyst, and pioneering computer scientist who helped break the Nazi code in WWII ("The Imitation Game" 2014), all the way to a composite fictional University of Chicago mathematician whose youthful genius has been replaced with deranged insanity and his brilliant and adoring daughter ("Proof" 2005). However complex their public presentations, the popular reception and consumption of these images in their visual representations embodies what I have tried to illustrate here in terms of the expectation of the

DOI: 10.1057/9781137519429.0004

benign, somewhat absent-minded professor, who is bound to accomplish great feats on behalf of humanity.

Against these idealized images of scientists socialized to pursue science for the love of it (or to explain divine mysteries, as Fuller suggests), we can then see the extent to which institutional failures are taking place. Given that science isn't a monolithic canon of theories and practices, but a more nuanced scientific community with multiple sub-communities and sub-disciplines, the terrain is more complex and opens up to institutionally sanctioned behavior that falls short of these ideals. For example, some laboratories expect more detailed accounting of one's materials than others, just as some journals expect more thorough compliance with conflict-of-interest disclosures than others. Moreover, scientific theories come into existence in many cases with the help of technological innovations, such as the telescope, rather than out of "pure" theoretical speculations or hypotheses. The historical and sociological interplay between science and technology and engineering, between the theoretical and the applied side of the equation has been so profound that the term "technoscience" has been coined since the 1970s (Sassower 2013a). Appreciating this interplay between science and technology is significant not only because we may wish to more accurately report on how science works or how the scientific community operates, but also because of the ethical dimension that this distinction brings to light. The classical notion that pure science is by definition value neutral whereas technology, as applied science, is necessarily fraught with moral quandaries doesn't work as well anymore. For example, who is responsible for the development and eventual deployment of the atomic bomb – theoretical physicists, engineers, the military brass, or the President? The shorthand designation of technoscience allows for a more nuanced and accurate way of examining the development and use of scientific knowledge in society, and thereby forces us to consider variables absent when considering so-called pure science alone.

As we move from *pure science* to *technoscience* as a set of activities that are informed by the technologies they use and bring about, we are inevitably moving outside the pristine confines of the universities into various sub-communities where other considerations become paramount. These communities are driven by political and financial interests as much as they are contextualized by social conventions and moral norms that might differ from those articulated earlier (as, for example, Merton's scientific ethos). Are we therefore better prepared to evaluate

DOI: 10.1057/9781137519429.0004

scientific practice in technoscientific terms? Are scientists able and willing to uphold rather than ignore the ethos of science while participating in the culture that surrounds them? Is it possible to shield the scientific community from external pressures so as to support and uphold its claims for an ideal ethos when this ethos is devalued and undermined?

Babbage's lament and warning

Before we turn to answer these questions in the next chapters, it might be helpful to examine one of the first warning shots that appeared to cross the scientific bow in the early 19th century. Though lamenting that "the pursuit of science does not, in England, constitute a distinct profession, as it does in many other countries [France and Germany]," the mathematician and philosopher Charles Babbage focused his attention on scientific pursuits and how to further their progress (2013/1830, 16). Fully aware of the fact that "long intervals frequently elapse between the discovery of new principles in science and their practical application," and that "great inventions of the age are not, with us at least, always produced in universities" (Ibid. 19–20), he considered the conditions under which the state should intervene and perhaps subsidize scientific research. According to him: "The question whether it is good policy in the government of a country to encourage science, is one which those who cultivate it are not perhaps the most unbiased judges. In England, those who have hitherto pursued science, have in general no very reasonable grounds of complaint; they knew, or should have known, that there was no demand for it, that it led to little honour, and to less profit" (Ibid. 21).

If neither the profit motive nor "honour" can secure the ongoing pursuit of science, it seems that only those who "possess a private fortune ... [and] can resolve to give up all intention of improving it" are the candidates for such a pursuit (Ibid. 29). Regardless of who will eventually join the scientific community and regardless of how the state will or will not support science, Babbage is the first (and the most quoted till today) to outline what he considers the four kinds or types of fraud found in science.

The first kind of fraud is "hoaxing," which includes complete fabrication of an event or fact – "no such animal exists." This is considered a "deception" that goes beyond the "strangeness" that an observation may yield (Ibid. 93), basically inventing something out of whole cloth.

DOI: 10.1057/9781137519429.0004

The second kind he terms "forging," which happens when someone "records an observation which he has never made" so as to "acquire a reputation" (Ibid. 94). Though he claims they are different, the first two kinds of deception illustrate similar intent to defraud by the means in which they are undertaken; in the former, putting together pieces that don't belong so as to claim a new finding/fact, and in the latter, simply inventing a fact that was never observed.

The third is labeled "trimming," which includes a process whereby the scientist is "clipping off little bits here and there." These trimmings may not change an average which is reported (cutting off the "outlying" observation as irrelevant or too far from the average), yet it is undertaken so as to "gain a reputation for extreme accuracy" (Ibid), when such accuracy is not present.

The fourth and last kind or type of fraud is called "cooking" the data which are reported by "select[ing] those only which agree" with or confirm a hypothesis (Ibid). This sounds like what we call "cooking the books" in bookkeeping and accounting, where revenue and expense are completely misrepresented so as to ensure profits or losses, depending on what the individual or corporation is expected to show shareholders or tax collectors.

The first two are more problematic but more easily discovered in short time when no other scientists report similar observations – which they cannot in principle and in practice (the "thing" doesn't exist!). The second two can be the result of either honest mistaken observations or errors in calculation, but when systematically practiced will eventually be found out by reconciling data notebooks and records with published presentation by referees and fellow scientists. Given the vocational expectations of Weber (personal integrity at the very least), those of Merton (as outlined by the four features of the ethos of science), or those repeated by Ziman (about the institutional normativity of science), any of these kinds of fraud are morally abhorrent.

Babbage, then, like those observers of the scientific community before and after him, reverts back to an idealized view of science. This time, his view is influenced by the speech delivered by Alexander von Humboldt which describes, in part, the ethos of science:

> The discovery of the truth without difference of opinion is unattainable, because the truth, in its greatest extent, can never be recognized by all, and at the same time. Each step, which seems to bring the explorer of nature nearer to his object, only carries him to the threshold of new labyrinths. The mass of

DOI: 10.1057/9781137519429.0004

doubt does not diminish, but spreads like a moving cloud over other and new fields; and whoever has called that a golden period, when difference of opinions, or, as some are accustomed to express it, the disputes of the learned, will be finished, has as imperfect a conception of the wants of science, and of its continued advancement, as a person who expects that the same opinions in geognosy, chemistry, or physiology, will be maintained for several centuries. (Ibid. 114)

Humboldt frames his view of the scientific community as a vibrant group of individuals who disagree with each other rather than a docile group of "normal" scientists in Kuhn's formulation. As the inquiry after the "truth" (a term used by Weber as well but curiously absent from many contemporary descriptions of science) will never bring about a "golden age" where consensus is reached, our culture should accept ongoing "disputes" and the continued construction of imaginative theories and the discovery of novel facts (Popper's conjectures and refutations, the emphasis on critical rationalism). This is what science is really about! Babbage was impressed by Humboldt because under this conception, science is an ongoing critical process where doubt and differences of opinions are standard rather than exceptional.

Babbage was hoping to enlighten the Royal Society of London with his studied critique of some of its members' failings, but instead of reverting back to a nostalgic view of the 17th-century gentlemen of science he draws attention to a German institution. Would fraud therefore be more easily detected under Humboldt's image of science? Would fraud be more readily swept under the rug when a (Kuhnian) consensus is expected or maintained within a paradigm? The hope was, and perhaps remains by some till today, that under continued "organized skepticism" (in Merton's words), scientists and the public at large would be more vigilant in examining all scientific knowledge claims and detecting fraudulent reports of the facts on which they are based. Perhaps here Greenberg's assessment that scientists "care deeply about their reputations," makes sense: their "shame" and "pride" might ignite a sense of personal and institutional integrity. This kind of integrity, this special function of scientists among us, has been understood in angelic terms, idealized beyond human frailty and weakness.

The image of scientists as fallen angels rather than angels at the service of the divine comes into play here. There is nothing more tragic than thinking of fallen angels at a deep existential level of reflection. Their fall is ours as well insofar as we have contributed to their loss of integrity:

DOI: 10.1057/9781137519429.0004

we are as guilty as they are in losing the whiff of divine inspiration and the quest for divine knowledge. If the biblical image of the Fall from the Garden of Eden is as much about eating the fruit from the Tree of Knowledge as it is about disobeying God's orders, then the image of fallen angels is as much about scientists losing their ways as about the way they have been positioned in the contemporary scientific enterprise. This enterprise is fraught with compromises scientists have to make to keep their jobs and grants, future funding and publications. Perhaps the biblical imagery, however foreign to most practicing scientists, remains operative in ways that inform scientists themselves and the publics they serve. Why expect of them more than they can deliver? Why put them on a pedestal they never consciously chose to occupy? Babbage's warnings about fraud in science may have been heard by some, but definitely not by all, as the next three chapters make clear.

DOI: 10.1057/9781137519429.0004

2
Big Science: Government Control of Science

Abstract: *In this chapter, the focus is on the means by which national interests compromise the autonomy of scientists to choose their research agendas and pursue their speculative inclinations. Whether under the guise of national security or patriotic pride, scientists have historically found themselves being controlled by the state apparatus. This has been true not only under dictatorial regimes but also within representative democracies.*

Sassower, Raphael, *Compromising the Ideals of Science*, Basingstoke: Palgrave Macmillan, 2015.
DOI: 10.1057/9781137519429.0005.

DOI: 10.1057/9781137519429.0005

One illuminating way in which scientists have compromised the ideals of science given the changing expectations of their work is when they have followed certain national ideologies or the leadership of their community (in Kuhn's sense) as has been observed in the Lysenko Affair in the Soviet Union and the Nazi Germany's hostility to "Jewish Science." Some of these ideological pressures undermine the autonomy of science as can be observed in large government-led programs, such as the Manhattan Project, Reagan's Star Wars, and even the Human Genome Project. In all of these cases, it doesn't matter that state oversight of the direction of the projects was for a higher cause or world peace, as they all illustrate the conditions under which the scientific community's expected autonomy is taken away.

Against the backdrop of idealized science – how the activities of scientific research are unlike any other human endeavor – there are certain political realities that became apparent in the 20th century (and continue into the 21st century). The very notion of science in the service of the state changes the expectation of scientists by challenging all that is dear to the ideology of the scientific community: universalism, communalism, disinterestedness, and organized skepticism. Once science is at the service of national interests, from security and defense to furthering the wellbeing of the citizenry, it is no longer science for science's sake, but instead science for the state's sake. This is true not only in cases of warfare, but also when nationally designated research priorities to cure cancer, for example, or fight contagious diseases are the items on the agenda. Once national authority takes over, however lofty and altruistic its goals and prospects, the scientific community loses its full autonomy and its ethos of independent research, no matter where it leads. The ideological underpinnings of this shift to national priorities in the name of national security, especially in light of two world wars, ferment a compromise of the ideals of science. In the name of patriotism, nationalist themes and pressures bring about a cognitive dissonance between one's identity as an idealized (and angelic) scientist and the definite result-driven expectations of funding agencies. This dissonance can foster feelings of doubt, guilt, and confusion, and these feelings characterize to some extent the mode of behavior in the era of Big Science. These psychological features translate into the institutionalized behavior of the scientific leadership. This argument will carry the first part of the chapter.

DOI: 10.1057/9781137519429.0005

Big Science in the service of the state

When scientific research requires resources no private or academic institution is able or willing to provide on its own, government coordination,
if not outright planning and administration, is required: this is what has
been called Big Science. Sometimes it's the scope of the project; at others
it's the vast resources associated with bringing it to fruition. Big Science
is usually associated with large-scale government-sponsored projects, the
ones against which President Eisenhower warned in his farewell address.
Alvin Weinberg, who was director of Oak Ridge National Laboratory,
argued that

> When history looks at the 20th century, she will see science and technology
> as its theme; she will find in the monuments of Big Science the huge rockets,
> the high-energy accelerators, the high flux research reactors – symbols of our
> time just as surely as she finds in Notre Dame a symbol of the Middle-Ages.
> She might even see analogies between our motivations for building these
> tools of giant science and the motivations of the church builders and the
> pyramid builders. We build our monuments in the name of scientific truth,
> they built theirs in the name of religious truth; we use our Big Science to add
> to our country's prestige, they used their churches for their cities' prestige;
> we build to placate what ex-President Eisenhower suggested could become
> a dominant scientific caste, they built to please the priests of Isis and Osiris.
> (1961, 161)

Already in 1961 Weinberg asks the scientific community to consider not
only the reasons behind the expansion of science into Big Science, but
also the price it was willing to pay for it. He asked three broad questions:
"first, Is Big Science ruining science?; second, Is Big Science ruining us
financially?; and third, Should we divert a larger part of our effort toward
scientific issues which bear more directly on human well-being than do
such Big-Science spectaculars as manned space travel and high-energy
physics" (Ibid.)? To be sure, just because there were national pressures
on the scientific community to participate in what was understood as
Big Science, it didn't mean that internal critiques, if not soul-searching,
weren't already underway.

One early American example of Big Science was the Manhattan
Project of World War II for the design and production of an atomic
bomb to fight fascism and bring an end to the war; another was the
sending of a manned spaceship to the moon after the Soviet Sputnik
program became successful. The story of the atomic bomb has been

DOI: 10.1057/9781137519429.0005

retold many times, with the excitement of mathematicians, theoretical physicists, and engineers poring over data and attempting a variety of experiments from 1939 to 1945. At various points, more than 130,000 people were involved at a cost of nearly $2 billion or about $26 billion adjusted to 2014. Research and production took place at more than 30 sites across the US, the UK, and Canada. Little Boy, a gun-type weapon, and the implosion-type Fat Man were used in the atomic bombings of Hiroshima (August 6, 1945) and Nagasaki (August 9, 1945), respectively, with civilian casualties in the range of 100,000–300,000 (depending on how they are counted). The Pacific theater of war came to a quick close after these bombs were dropped and nuclear weaponry proliferated. These are the bare facts of the matter, facts that bring to light how the state controlled science under conditions of war. This is not to suggest that there weren't then nor that there aren't till today many conflicting arguments about the morality of this project from its inception to its devastating results of killing civilians. Likewise, it should be mentioned in passing that objections to and critiques of the development of nuclear warfare devices became more vocal once the bombs were dropped on Hiroshima and Nagasaki. This was true all the way from Einstein writing letters and articles, Robert Oppenheimer's (the leader of the Manhattan Project) famous plea with President Truman, to the formation of the Union of Concerned Scientists and much later organized groups, such as the International Physicians for the Prevention of Nuclear War.

In light of redrawing the European territorial boundaries between the Allies and the Soviet Union after the end of World War II, the Cold War was in full force. It was important for the Americans to win that war, if not on the battlefield then at least in the imagination of its citizens. The National Aeronautics and Space Administration was founded in 1958 (re-baptizing NACA) and became responsible for the nation's civilian space program and for aeronautics and aerospace research. On July 20, 1969, Apollo 11 was the spacecraft that landed the first humans on the moon, Neil Armstrong and Buzz Aldrin, as a direct consequence of President Kennedy's challenge to the American technoscientific community. The cost of the entire Apollo program was $25.4 billion or $135 billion adjusted to 2005.

More recent Big Science projects include the Reagan Administration's Strategic Defense Initiative (SDI) also dubbed by critics as "Star Wars," at an estimated cost of over $200 billion (from 1983 to the present). Reagan's ideology was hawkish enough to ignore international treaties

DOI: 10.1057/9781137519429.0005

and push the development of ground-based and space-based systems to protect the US from foreign attack by strategic nuclear ballistic missiles (Broad 1983). The initiative focused on strategic defense rather than the prior strategic offense doctrine of Mutual Assured Destruction (MAD). The Strategic Defense Initiative Organization (SDIO) was set up in 1984 within the US Department of Defense to oversee SDI. Under the Clinton Administration, its name was changed in 1993 to the Ballistic Missile Defense Organization (BMDO) and its emphasis was shifted from national missile defense to theater missile defense, and its scope from global to more regional coverage. BMDO was renamed the Missile Defense Agency in 2002.

A third example of a Big Science project was the Human Genome Project which was conceived by the US Department of Energy (DOE), took 13 years to complete (planned in 1984, began in 1990, and completed in 2003), and was coordinated by the DOE and the National Institutes of Health (NIH). It sequenced 3 billion base pairs of human DNA, identifying and mapping all of the genes of the human genome from both physical and functional standpoints. Most of the government-sponsored sequencing was performed in universities and research centers in the US, UK, France, Germany, Japan, Spain, and China. The cost was around $3 billion. This, of course, is Big Science by any definition, from the amount of funding required to accomplish its goals to the number of scientists and institutions that coordinated their efforts to successfully complete the task. It should be noted that a parallel effort by Celera Genomics began in 1998 (a private corporation whose motive was commercial in addition to the expansion of human knowledge), and according to some was a helpful competitive motivator for the larger government-sponsored project to continue its own effort.

This case also illustrates how difficult it is to draw clear lines of demarcation between state Big Science and commercial Big Science. In both cases, one has to consider the obvious concerns over intellectual property that arise when either public or private enterprise undertakes basic research projects (more on this later): should private enterprises compete with government agencies? Should the one take risks the other cannot afford? Is the perception indeed true that the private sector is necessarily more efficient in its use of funds than the public? The difference between the two competing human genome sequencing projects, according to Philip Mirowski, is that whereas the government-sponsored research is open to the public (in terms of free access to information),

DOI: 10.1057/9781137519429.0005

the private-sector project could keep its findings as proprietary secrets even when some results were published in professional journals (2011, 292). The initial intent of the research makes a difference in what evolves later: is there public pressure to find more about genetics? Or, is it by contrast the expectation of pharmaceutical companies to capitalize on this research through the development and sale of new therapies and drugs? Citizens in democratic states can vote with their wallets, so to speak, and demand from their representatives more research funding in one area or another. But however sophisticated their own understanding of science and of the political system that can bring about such legislative action, they will always fall short of the quick, decisive, and financially powerful actions of a few entrepreneurial individuals or corporations who set targets and can muscle the financial resources to fund them.

Though World War II and the Cold War defined Big Science in the 20th century, issues relating to how a nation should or should not deal with science and its community has been with us since the days of Babbage (as seen at the end of Chapter 1). Should the state fund science? If yes, should it intervene in its direction and expect certain outcomes? In broader terms, should the ideological commitments of the state guide or determine the activities of scientists? After the Bolshevik Revolution of 1917 and the total control of the Soviet state by the Communist Party, it was becoming obvious that scientists, like peasants and farmers, journalists and artists, must heed the party line. Its ideology, as expounded by Vladimir Lenin (1870–1924), insisted on science coming to terms with Marxist dialectical materialism. This meant not only a classless society where the proletariat would forge ahead a promising new future, but also the ability of the state to form a new social order (and even tamper with human nature) that conforms to this ideology. Within this context, some scientists were considered "bourgeois" and as such were deemed dangerous to the progress of the state. Some were dismissed from their positions, some sent to the gulags, whereas those willing to participate in the revolutionary activities of the state remained in their posts, eventually promoted beyond their levels of competence.

Among the most notorious of these scientists was Trofim Lysenko (1898–1976), who was supported by Joseph Stalin (1878–1953) with devastating results. In general, there are two genetic schools of thought, the one represented by Jean-Baptiste Lamarck (1744–1829) and the other by Gregor Mendel (1822–1884). Whereas Lamarck claimed the evolutionary process to be working through individual plants (or humans)

DOI: 10.1057/9781137519429.0005

adapting to environmental conditions during their lifetime and (if they survive) passing these mutations to the next generation, Mendel focused on random genetic mutations as the driving force of evolution. The up and coming Soviet geneticists were rebuffed by those more concerned for agricultural research and the provision of food to the population, those for whom science was to be at the service of practical concerns. David Joravsky (1970) recounts what has been dubbed the Lysenko Affair as "thirty-five years of brutal irrationality" (vii). At stake for Lenin and eventually Stalin was the erasure of a long-standing compromise with scientists to distinguish between their political and scientific views, the former being under the control of the Party whereas the latter remaining autonomous and protected from ideological pressures (Ibid. 37).

Though there seemed to be some initial success in Lysenko's experiments with increasing crop yields with the use of "vernalization" (soaking seeds in water before sowing them), it was a success limited to half a hectare on his father's farm. As such, it was an interesting experiment, but one that in order to be claimed as scientific in any sense of the term discussed earlier would have been questioned, criticized, and perhaps relegated to the dustbin of history. Promoting "hut labs" and "peasant scientists," Lysenko and Party officials could dispense with the scientific institutes of the day and claim the authority of science (Ibid. 90). Instead of adhering to the universal scientific standards of the day, a great deal of public relations effort on behalf of this new "agrobiology" was promoted by Lysenko himself and hordes of Party officials pressured to provide practical solutions to the declining production of grains in the USSR (Ibid. 60–61). Without dealing with the sham details and pseudo-proofs he provided in his Moscow Institute of Genetics, what the Lysenko Affair exemplifies is the manner by which political power overwhelms the strictures and ideals of science so as to dismiss challenges – repeated testing and confirmation or falsification – on behalf of a state ideology.

Lysenko played an important political and even practical role for the Stalin regime with its own agenda rather than remain true to any part of the scientific ethos. The ideology that motivated this unscientific way of thinking was not only influential, but it also undermined the ethos of science by the state protection of pseudoscience. Though several critics remained, especially among the geneticists who opposed the Lysenko group (but who could be labeled "bourgeois" and thus discredited), on the whole the Stalinist fusion of theory and practice (where practice is the motivation for research and the final arbiter of what counts as

DOI: 10.1057/9781137519429.0005

scientific) found its best mouthpiece in Lysenko who not only proposed sham practices under the guise of science but also worked diligently to discredit the existing scientific community (Ibid. 83). As we shall see later in the case of the dominance of business interests over scientific integrity, here, too, scientific principles and methods could be misused: when Lysenko wanted "proof" he obtained it, no matter how small his sample or experiment was; when he couldn't obtain such proof of his "scientific discoveries" he promised great gains in the future, thus postponing potential failures or external examination (Ibid. 111). The history and methodology of science provided plenty of opportunities for Lysenko's subversion and manipulation; and he was a willing and subservient participant in the control of the state over science.

A similar mindset overtook the Nazi regime in Germany before and during World War II, so that anything considered "Jewish Science" was deemed false because of its crypto-racist overtones (Fuller 2004, 104). Francis Galton's classical ideas of eugenics were revived with an ideological zeal unparalleled in the history of German science. This nationalist fervor was in part a reaction to British physicists accepting their own national propaganda about German atrocities of World War I. With proper party guidance, Aryan supremacy could be "scientifically" proven, and any foreign challenges jettisoned. No different from Hitler's aversion to modern art of the early 20th century, deemed "Degenerative Art" and condemned to be banned from museums and galleries if not burned, "Jewish Science" was likewise considered subversive, dangerous, and profoundly illegitimate. Every facet of the scientific ethos we encountered in the previous chapter was ignored in order to promote a political view that needed to be scientifically credible and pure rather than simply ideologically tainted.

Outright ideological subversion of science of the Soviet or Nazi kind is easy to detect and discredit in retrospect. As Vadim Birstein (2001) persuasively argues (with the advantage of consulting secret files in closed archives that became available for public scrutiny only after the Berlin Wall came down and the Soviet regime collapsed), is that "the perversion of knowledge" (the book's title) demanded a compromise of conscience by the scientific community. Raising moral concerns, Birstein has a scathing critique of the ways and means used by fascist regimes in the 20th century that eviscerated the integrity of the scientific community. It's not simply that we eventually observe the silliness of insisting that something is scientific when it's blatantly not the case, or when

DOI: 10.1057/9781137519429.0005

something is relegated to the pseudoscientific bin when it's in fact thoroughly valid. Nor is it the case that we might be confounded by moral lapses of judgment. Much more difficult to ascertain when it happens is the ideological underpinnings that subvert what seems to be taking place in a free democracy. Under these conditions of apparent openness, a more subtle but no less profound influence can still be detected when we are alert enough to notice it. According to Greenberg, "science in the American polity is well supported and respected, despite its habitual assertions of neglect and hostility, but its most influential institutions are underscrutinized by other sectors of society" (2001, 2). This doesn't mean that there is no scrutiny whatsoever, but that there is a deferential kind of scrutiny. The insulation of science from the rest of society is partially an ideological commitment to the unhampered pursuit of scientific ideas, as if the state has nothing to do with it. As science sets itself up as a religion of sorts, where "true believers are fervent in their faith and uncomfortable with or even hostile to external scrutiny," the claim of self-policing is akin to a demand that no one should ever challenge matters of faith. This means that "outsiders, at least in tolerant lands, politely refrain from critical inquiry and commentary that might be regarded as disrespectful" (Ibid). With this in mind, science receives a free pass from those who fund it; a moral duty to fund science is reciprocated with the moral duty to accept whatever scientists say as if they are speaking, using the Catholic term, "ex-cathedra."

Vannevar Bush, more than anyone before or after him, was responsible for harnessing an ideological commitment to science different from the Soviet and Nazi models primarily because instead of asking "what can science do for the state?" he asked "what can the state do for science (that would be eventually worth the investment)?" He was head of the Office of Scientific Research and Development (OSRD) during World War II, through which almost all wartime military R&D was carried out, including initiation and early administration of the Manhattan Project; he was also the Chairman of the National Advisory Committee for Aeronautics (NACA), and the Chairman of the National Defense Research Committee (NDRC); he coordinated the activities of some 6000 leading American scientists in the application of science to warfare. As the first presidential science advisor he was a "science politician" who was able to formulate the benefits of government investments to elected politicians who were basically scientifically illiterates. He also was instrumental in perpetuating the myth of "freedom of inquiry" and of "free play of free intellects,

DOI: 10.1057/9781137519429.0005

working on subjects of their own choice, in the manner dictated by their own curiosity for the explanation of the unknown" (Ibid. 52). Exploiting the religious imagery that dominates American culture, it was natural to perpetuate "the dual myth of Bush as messiah and *Science, The Endless Frontier* as sacred text" all the way to the present (Ibid 55; italics denote his book's title).

The insulation of the scientific community from the pressures of the state may seem the perfect antidote to the kind of ideological interventions and subversions of the Soviet or German seen before; however, with the *linear model* of the *scientific enterprise* the state ends up supporting a specific agenda of monumental ideological importance: fight fascism (WWII) or the "evil empire" (USSR during the Cold War) with the valuable results of applied scientific research (Fuller 2004, 27). This means that though the state seems removed from direct control of its scientists, it nevertheless exerts an enormous influence on them and the ways they must conduct their research. The linear model starts with basic research that is funded by the state and ends with technologies that benefit the state apparatus. After the Cold War ended in 1989 the new beneficiaries were the markets, extending the political framework to enhance the activities of capitalist markets. Al Gore, as US Senator and Vice President, tried to insert environmentalism in addition to other state projects and goals, but failed. In other words, what seems a reasonable way of protecting the ideals of science, ensuring that though originally funded by the state the scientific community may develop its own research agendas, turns out not to be so. Instead, the state provides the conditions for private companies (as a form of outsourcing rather than directly supervising scientific projects) to capitalize on its investments in basic research, and then exploits them in ways that compromise the independence of the scientific community. Perhaps the dictates of the state are now subtly (or not so subtly) transformed into those of the marketplace.

What is important to appreciate in light of the ideological concerns posed by government support for scientific research is that just as much as national interests direct the scientific community, they are also indicators in the modern world of how much nations care about scientific progress. Perhaps the key to the success of the scientific enterprise – now fully understood in terms of the financial backing it receives from the state (and corporate sources funded indirectly by the state) – is also the disconnect between the public understanding of science and the knowledge claims made by scientists. This is also true of the different priorities

DOI: 10.1057/9781137519429.0005

the public may have in mind – more funding for cancer research than particle physics – as opposed to the scientific community. This disengagement or separation between the public at large and the scientific community isn't simply predicated on the lack of public understanding of and interest in science, but is also indirectly encouraged by the scientific community itself. The scientific leadership is worried by overly engaged politicians representing public sentiment because of the fads and fashions they may follow, instead of them having a clear agenda that may take decades to explore. Scientific leaders ask for funding with an expectation of free reign and little oversight, whereas the public expects in return that technoscientific research will be in the public's best interest and respond to its needs.

Though some would argue that the controversies over Star Wars were the last major eruption of political passion in the scientific community – in terms of the military use of satellites in what was considered a territory outside of the reach of terrestrial warfare – they were reminiscent of the scientific promises and horrors of past wars. One can supplement this claim by arguing that the controversies over AIDS research (to be discussed later) also brought about a public outcry and gave rise to political passions. But perhaps there is a point here about the turbulent relationship between scientists and the public, insofar as some guilt-ridden scientists (under public moral questioning) wished to shift the focus from military buildups to the broader and more peaceful potential of scientific research, especially for health care and the environment. Perhaps what replaced the obsession with funding warfare or the means by which "national security" is protected, has been the shift of government-sponsored research (as basic as it gets rather than only its applied aspects) to outright licensing of DOD research for private purposes.

As Mariana Mazzucato brilliantly documents, the state has taken immense risks in funding basic research – whether originally understood in warfare terms or national security ones – whereas private enterprise has reaped the rewards with minimal licensing fees and at no risk whatsoever. This two-step process – state research and private corporate licensing – makes sense if it's designed from its inception to invest publicly in basic research so that individuals and companies can eventually benefit from this common pool of research – a *research commons* of sorts. If the state becomes the clearinghouse of ingenuity, if it funds and collects the best resources no individual or company ever could, then it makes sense for it to offer its collective knowledge after the fact to

DOI: 10.1057/9781137519429.0005

anyone who can make use of it (for a fee, of course). Mazzucato argues that "it was the visible hand of the State which made these innovations happen [iPhone, SIRI, GPS, touch-screen]" (2014, 3). This *visible hand* (rather than the so-called *invisible hand* of market capitalism) is a significant promoter of Big Science, and one which determines the behavior of the scientific community and its ability to adhere to the ideals of science. But, as argued later, the realities of state entrepreneurship are ignored by the neoliberal argument about the inefficiencies and bureaucratic perils of the state and the efficiencies of the marketplace.

The effects of market capitalism

The first kind of compromises scientists undertake given the changing expectations of them denotes a shift from universal, communalist, disinterested, and an institutionalized skeptic ethos to a nationally informed ethos of patriotic zeal that mixes warfare mentality with national pride. As such, this change of expectations of the ideals of science focuses on the shift from a self-generated research program of the scientific community to one dictated from the outside, especially because the power of the public or private purse is unmistaken. Without surveying different public policies according to which funding allocation takes place – how much should be devoted to medicine as opposed to physics? – it becomes apparent that state control may have a corrupting effect not only on the choice of research but also on the institutional organization of the scientific community (as we have seen earlier).

A second kind of compromises scientists undertake in light of the changing expectations of them is still bound by the state apparatus, but instead of being informed by its nationalistic tendencies, it's informed by its adherence to the capitalist ethos of regulation-free markets. When democratic ideals of freedom, equality, representative government, the rule of law, and majority rule are subverted and appropriated to conform to market forces (and eventually to what some have started labeling "super-capitalism" or "hyper-capitalism"), then these ideals are not simply compromised but in fact disappear. Scientific research is understood in this phase as an enterprise, where financial resources and potential profits are paramount while expecting to be shielded from government intervention. Rather than insisting that the scientific community "return" to an idealized state (that was only imagined but

DOI: 10.1057/9781137519429.0005

never existed, as explained earlier), this ideological shift uses the rhetoric of neoliberal ideology to jettison the control of the state (and all its regulatory agencies) in favor of entrepreneurial corporations whose pursuits (for fame or profit) are ideologically sanctioned and legally protected.

Big Science, though, hasn't gone away, as large projects still require planning and coordination, enormous amounts of funding and clearly identified goals. When the scientific community is expected to behave like an economic enterprise, still demanding state funding but refusing its control, and expecting complete independence while shirking public transparency, the framing of its activities has radically changed. What meaning does the *scientific commons* – just like other public goods that are funded by and for the public – have under these conditions? Can the scientific community remain true to its ethos and work in the public's best interest, while ensuring profitable products to its funding agencies (private and public alike)? This quandary could lead, as seen later, to a series of compromises and fraudulent abuses.

As the case of warfare research or scientific research for warfare purposes illustrates, though presuming freedom of inquiry, there is a set of goals set by government agencies that fund the scientific community. Though not always explicitly stated, this means that the scientific enterprise – as it has become in light of the dominance of financial needs for its administration – is explicitly directed into certain areas of research rather than others. Whether understood in terms of national security or in terms of providing the tools for the promotion of national interests, big government and corporate funding ends up dictating the direction of scientific research even when losing control over its operations.

Though regulation developed in the US already in the mid-19th century as democratic control and service to citizens who were worried about the unbridled excesses of individuals and the companies for which they worked, the very notion of regulation or Big Brother watching over one's activities is squarely associated by the mid-20th century with the totalitarian regime of the Soviet Communist Party and therefore treated with suspicion and hostility. There is some confusion and debate over whether regulation is meant to solely provide the legal framework within which we operate (providing protection for health and the environment) or rather to constrain one's ability to perform the complex tasks of technoscience. In this ideological climate of the free pursuit of scientific inquiry (with government funding), independent conduct is expected. The standard practice of oversight is thereby challenged, and without

DOI: 10.1057/9781137519429.0005

oversight, of course, abuses can more easily be exercised (Bowker 2003, 87–88). The tension between too much or too little regulation cannot be resolved by the courts alone, even though they can provide clarity of perspective when muddled arguments are offered. Instead, it must be daily renegotiated by all the parties involved in research and development, recognizing changes in the circumstances under which scientists work.

The anti-regulatory instinct was bolstered by what Bush codified as the new, post-World War II ideology of science: "Progress in the war against disease results from discoveries in remote and unexpected fields of medicine and the underlying sciences." This means that the potential for scientific progress depends on the "free play of free intellects," as already mentioned earlier (Quoted in Judson 2004, 22–23; italics in the original). With their demand for absolute freedom of thought and practice, scientists can have their cake and eat it, too: accepting funding from government agencies while refusing to meet specific expectations. This doesn't necessarily imply that they really believe in "anything goes" kind of ideology or that they refuse to monitor themselves or police their fellow scientists. They continue to serve as reviewers of projects other than their own, and they continue to publish their results in professional peer-reviewed journals. It only means that the quest for autonomy is no longer framed in ideological or philosophical terms, but instead is framed within specific public or private funding conditions (Agassi 2003).

Facing such an attitude by the scientific community, why should governments (representing tax-paying citizens who must continuously fill national coffers) support science? Perhaps what is at stake is not only military prowess and national security, but also a sense that national pride and status in the world community is easily measured in terms of scientific progress. Scientific distinction is a clear symbol for economic growth and sustainability, improved health care and environmental stewardship. By contrast, less-developed nations are commonly characterized in terms of poor support for scientific inquiry and technoscientific backwardness that hampers economic growth. The scientific leadership rightly expects a certain latitude in its research programs, while wise politicians allow for such latitude with an expectation that such an investment will yield successful results (Grant 2007, 269–270).

Theoretically, politicians should keep their hands off the controls of the scientific brain train; practically, though, they cannot resist the temptation to take credit for achieving tangible goals for their constituents.

DOI: 10.1057/9781137519429.0005

The rationale for state control over science isn't limited to accountability (how are you spending our constituents' tax dollars?), but is associated with the so-called linear model mentioned earlier. It has the following schema: Federal funding → Basic research → Applied research → Development → Technology → Application → Social benefits. Though originating much earlier, the fervor of warfare mentality catapulted this model which has persisted through the Cold War and remains the yardstick (if not an actual representation) of how to manage peaceful deterrence projects without fully transforming funding priorities of the DOD. The withdrawal of the state from its role as science patron and manager may have been a welcome change in attitude after 1989 (can it pick winners?), but in point of fact it still remains a major donor of all scientific projects in US universities (notwithstanding corporate R&D which some estimate to be twice that much): "In 2005 universities spent $45.8 billion on R&D. Of that amount, the federal government provided $29.2 billion" (Greenberg 2007, 39). According to Mazzucato, it's not only the amount of state investments that is staggering, but that "in biotechnology, nanotechnology and the Internet, venture capital arrived 15–20 years *after* the most important investments were made by public sector funds" (2014, 23; italics in the original).

Unlike the "pro-statist position" that favored the linear model, the neoliberals who came to prominence in the Reagan and Thatcher Administrations wanted to privatize every step of this model, from funding to application, so that the market of ideas had a for-profit telos. Neoliberals (of the Friedrich Hayek (1899–1992) and Ludwig von Mises (1881–1973) tradition) lost their cause when "public goods" (or the scientific commons) arguments came about during World War II and the Cold War; but after the 1980s, especially with the fall of the Berlin Wall, they regained dominance (Mirowski 47–60). The neoliberal attack was based on two principles: "(1) there was no such animal as public good, once you looked at things properly, and (2) all knowledge was always and everywhere adequately organized and allocated by markets, because the market was really just one super information processor" (Ibid 61). The real issue is who owns government-financed research, pitting "populists" who argue about the scientific commons that the public owns, against "free-enterprisers," for whom ownership remains in the hands of corporations (Greenberg 2007, 53). A contributing factor that tipped the scale toward a neoliberal interpretation is the Bayh-Dole Act (Trade and Patent Amendments Act of 1980) that "imposed on them

DOI: 10.1057/9781137519429.0005

and their scientists a duty to pursue licensing to industry as a condition of accepting government money for research" (Ibid. 54). With this legislative edict in mind, it's quite clear how and why private corporations end up directly or indirectly owning the research produced with public funding and its fruits.

Under these congressionally sanctioned conditions of market-allocation efficiency or more accurately, armed with these assumptions, neoliberals looked at the linear model with outright contempt, attempting to reorient its various steps wherever possible. One can only wonder why any military-funded science remained classified, or became "gray" or "black" or "secret" science all the way into the present (Ibid. 111). This isn't meant as a way to protect national interests, but to become a vehicle to subvert the scientific ethos. It's also a way to ensure that certain "fruits" of this publicly funded research can be picked at low licensing prices by private industry, once its efficacy is proven by military application (as was the case of the Global Positioning System). As odd as it may sound, the interests of the state to keep scientific knowledge secret mesh with capitalist interests to codify and secure intellectual property rights regardless of their original funding. This may be more of a US-centric view, as some European attitudes are quite different. Nationalist and capitalist interests coalesce to ward off public scrutiny and challenge the idealized vestiges of the democratic scientific ethos.

There is, of course, another critical trajectory that conservative, neoliberal market capitalists use when considering government funding of scientific research (or for that matter, any government funding whatsoever). Though this second kind of critical trajectory poses more objections to government bureaucracies than to the structure and administration of scientific community, the two targets seem to be lumped together so as to equally enjoy the wrath of the critics: they are all wastefully sponging off the public trough. According to Joseph Martino, government funding of science "has also resulted in political criteria for support of scientists, congressional micromanagement, the freezing out of innovative ideas, the favoring of big science over little science, and pork barrel science. Ultimately, it will lead to the corruption of the American scientific enterprise. Moreover, these evils are inherent in government funding" (1992, 371). For conservative critics like him, "the distortion and corruption were not things that could be corrected by replacing one set of bureaucrats by another. They were inherent in federal funding and are present in every other enterprise funded from

DOI: 10.1057/9781137519429.0005

the federal treasury" (Ibid. 385). On the verge of fear mongering, this extreme rhetorical expression is quite common among conservatives in our political spectrum, and can be witnessed in popular media outlets, from television networks to Internet blogs.

Finally, there is a third kind of criticism of how institutional structures have conformed to (state-sanctioned) business interests by exploiting the regulatory apparatus they themselves attempt to avoid. The paradoxical posture is astonishing as much as it is effective, as Robert Proctor cleverly notes: there are various rhetorical strategies used by private corporations to avoid either regulatory scrutiny or admitting that they are culpable for any public hazards. Among them, he cites the following language: "Evidence of toxic hazards is ambiguous, inconclusive, or incomplete; we therefore need 'more research' to clarify ambiguities, improve estimates of risk, elucidate mechanisms, and so forth" (1995, 125). This can be considered as a sound scientific response, methodologically respectable, and communally responsible. But a call for "more research often translates into a call for 'less action' – and, specifically, less regulation" (Ibid. 130). In this context, scientific uncertainty and the modesty of scientific claims – hallmarks of the institutionalized skepticism that are an integral part of the ideals of science – are being subverted. In other words, instead of truly worrying about making too outlandish claims, the purported claims of uncertainty are exploited for the sake of shielding corporate maleficence rather than improving the conditions under which toxic hazards can be clearly identified and denounced.

For Proctor, the political climate of deregulation is not limited to the Reagan and Thatcher years (as Mirowski agrees), but has continued into the Clinton and Bush Administrations. It is also not limited to government agencies and their direct control or administration of scientific projects. Instead, this anti-regulation ethos filters all the way down to professional and trade organizations that are appropriating the rhetoric of the scientific ethos, not in order to incorporate it into their corporate members' practices, but on the contrary, in order not to fulfill the promises of science. In his words: "Science plays an underappreciated ideological role in the industrial context. Trade associations appeal to science to give themselves a semblance of neutrality, balance, and level-headedness in questions of environmental risk assessment. The appeal to science creates an impression that the company is making progress, keeping up with the times. It also buys time. The call for more research

DOI: 10.1057/9781137519429.0005

often works to delay implementation of regulatory standards – paralysis by analysis, as a 1980 OSHA document put it" (Ibid. 131).

It may not be clear under these circumstances where guilt lies: Is it with the scientists who work for their corporate paymasters or with their masters alone? Scientists are complicit when they let their paymasters present their findings in legally couched and misleading terms rather than directly addressing the public with plain empirical facts. Given that so much of corporate research is funded by government agencies – either with direct disbursements or indirectly through tax credits and incentives (as will be shown in later chapters about pharmaceuticals), it makes sense already in this chapter to identify the changing expectations of scientists associated with state-sponsored research, and the regulatory apparatus that surrounds its largess. In other words, scientists find themselves making compromises they didn't expect to be making when they first undertook their research, expecting that once the boundaries of their research were outlined they could maintain their professional integrity.

Proctor represents a view that is concerned with several kinds of corruption and the ways they have changed the conditions under which scientists operate, conditions to which at times they themselves contribute. In addition to the *first* tactic of "paralysis by analysis" (delay tactics to avoid compliance with regulatory standards), there are *additional diversion* tactics that shift public attention (and that of regulatory agencies) from "questions of health" to "questions of free speech." In such cases, a debate over pollution or toxic side effects is preempted by a discussion of allowing alternative voices to be heard about the pros and cons of pollution even if these voices' direct relevance is contestable. A *third* set of tactics is used by corporate scientists and their spokespersons who rely on the scientific insistence for thorough empirical verification. In this case, "the rabid distrust of a well-founded hypothetical can be used to dismiss a plausible point – as when animal evidence of carcinogenicity is dismissed by industry-financed skeptics." These tactics, too, delay action and can provide cover for months and years until proper findings are universally accepted. Similarly, in the *fourth* tactic that is grounded in the methods of scientific inquiry, "a misplaced call for precision can also be wielded as a political tool – as when calls are made to delay regulation until endlessly more elaborate studies confirm that a substance is hazardous." The case for precision in the name of reducing uncertainty and doubt can of course become a device for endless research, as full

DOI: 10.1057/9781137519429.0005

certainty is never achievable even under the best of experimental conditions (Ibid. 131–132). All four tactical sets can be used individually or together; they can be deployed at any time with a clear conscience that science is served best when investigations take their time, when all the evidence is meticulously collected, and when doubt is replaced with certainty.

With all of these tactics in full use, there is no doubt that some scientists, directly or indirectly, wilfully or under duress, are feeling the changing expectations of their practice and the compromises the ideals of science have suffered over the years. The perverse use of scientific ideals on behalf of corporate interests – and with the full endorsement of the state, its regulatory agencies, and its legal system as a whole – is in full view in the halls of democracy. To be clear, the alleged full endorsement of the democratic state of such misconduct would suggest that the state is representing the majority wishes of its citizens, but of course one would be horrified to think that the public doesn't want to be protected from potential hazards. One must question what it means for the majority to be "represented" in this case. Clever and seemingly responsible, this form of the corruption of science makes a mockery of the ideals of science and exposes the realities of the scientific enterprise, pressured as it is to hide or overlook potential and actual side effects or dangerous unintended consequences of technoscientific applications. Against this background, we should examine some preventive interventions put in place in the US.

National guidelines for scientific integrity

The chapter ends with some guidelines that have been adopted by various government agencies to regulate science and prevent fraud (of course, they are expected to be unlike state-control edicts in totalitarian regimes where sheer control overshadows any and all reasonable considerations of the free spirit of science). The first thing that resurfaces at this juncture is the idealized vision of scientific conduct. The worry over the compromises scientists have had to make to accommodate the changing expectations of their institutionalized working conditions is understood here in light of Weber's notion of "vocation" that implied total commitment to science, the acceptance of change, the knowledge that one's work will be superseded, and the ongoing demystification of nature by rational means. The shift in expectations or the compromising process has proceeded from the Weberian heights to the Kuhnian lows, where vocation is transformed into puzzle-solving, routinized professional activity, and adherence to the leadership of the scientific community. If we are dealing with a socialized workplace, there is no wonder that we'll find fraud as fabrication, falsification, and plagiarism, but these broad definitions have different meanings if understood according to the scientific ethos or the legal system; it's unclear that both sets of definitions are indeed commensurable. And as we examine some of them, it'll become clear where the limits of the legal definition leaves much room for the more professionally responsible one.

The National Science Foundation's definition of fraudulent activity is as follows: "(1) Fabrication, falsification, plagiarism or other serious deviation from accepted practices in proposing, carrying out, or reporting results from activities funded by NSF; or (2) Retaliation of any kind against a person who reported or provided information about suspected or alleged misconduct and who has not acted in bad faith." By contrast, the Public Health Service's definition is as follows: "'Misconduct' or 'Misconduct in Science' means fabrication, falsification, plagiarism, or other practices that seriously deviate from those that are commonly accepted within the scientific community for proposing, conducting, and reposting research. It does not include honest error or honest differences in interpretations or judgments of data" (Quoted in Judson 2004, 172). How do these definitions reformulate Babbage's classification of "hoaxing, forging, trimming, and cooking?" Moreover, when the legal system is brought up, we should ask, are we dealing with criminal or civil offenses? Should the standard of the Internal Revenue Service be used, for example, where innocence is assumed until fraud is proven, and once proven, it's presumed to be "intentional" (Ibid. 384)? Should court of law judges be the "gate-keepers" when it gets to scientific evidence, using the same criteria they use in civil and criminal cases? According to the Supreme Court, lower court judges are entitled to ask for independent panels to assess expert testimony. But when one set of experts challenges another, will this legal process lead to infinite regress, where appeals are heaped one on another, and where any set of experts can find flaws and doubt in the other's testimony? Is this legal process a preferable system to the European one where Continental judges are entrusted with the role of "examiners-in-chief" rather than that of "mediators" of evidential disputes (Ibid. 386ff)?

DOI: 10.1057/9781137519429.0005

No matter what criteria are offered by courts of law or government agencies, no matter what process is followed, they are obviously set up in response to an outcry over scientific fraud or unpopular scientific research. Judson moves back and forth from assigning blame to individuals through the many case studies he reviews, to thinking about institutional complicity as the driver of these fraudulent activities. He, like others, speaks of the young research prodigy, the mentor who is being seduced, and the pathologies that mutually infect them: complicity of co-authors, and the arrogance of power, where one commonly finds "willful blindness," "failure to supervise," and "failure to communicate" (Ibid. Ch. 3). Judson looks for generic and diagnostic traits that permeate his case studies, and therefore he talks about "strong, productive, demanding mentor with an empire and little time for supervision ... [and] a younger scientist of brilliance, charm, and plausibility with a record of publication and exciting research elsewhere ... [and a] gift authorship [that] leads to failure to detect error and is itself misrepresentation" (Ibid. 144–145). This description allows readers and analysts to identify with such situations, to recall when and if they themselves were party to this psycho-social dynamics within a laboratory or graduate school context (Latour and Woolgar 1979). How do we perceive brilliance? How do we interact with our mentors and mentees? Should we question those who offer us a free ride with a publication? Should we be suspicious of any co-authorship? Perhaps the answer to all of these questions, with an eye to the ideals of science, should be in the affirmative.

These questions and the conditions that bring them about at once blurs the distinction between personal and institutional complicity, because routine and repeated personal interactions become standards for the community of scientists as a whole. Perhaps the competitive nature of grant submission and awards underlies all of these potential conflicts and is at the heart of the problem discussed here. Perhaps what we are witnessing by the 21st century is both structural ("rotten-system theory") and personal (curiosity vs. fame and fortune) (Ibid. 146). His conclusion is that "fraud and its consequences cannot be understood except by addressing the social dynamics of laboratory life, and then their setting in the larger and evolving communities of the enterprise of science" (Ibid. 150). But then he also says, in contradistinction to his conclusion, that "fabricators and falsifiers in the sciences originate their frauds alone" (Ibid. 312). Which is it, then? We move back and forth from assigning guilt to individuals and to the laboratories or institutions

DOI: 10.1057/9781137519429.0005

in which they are employed; we think that the guilt of the few reflects poorly on their environment, while believing that their environments are understood as the amalgam of the behavior of individuals.

Beyond the cases and numbers associated with fraud, there is also the mindset, the rhetoric used in cases of scientific fraud. If we agree, as Howard Schachman correctly insists, that "brilliant, creative, pioneering research often deviates from that commonly accepted within the scientific community," on whom should we place the burden of proof or blame (1993, 150)? Sometimes what may seem at first to be unreasonable, even suspicious piece of experimental evidence, may turn out much later to be much more acceptable and reasonable experimental result. Has the change of terminology from "fraud in science" to "misconduct in science" broadened the net for abuse? Or has it softened the language so that the legal force of courts of law is curtailed, allowing more cases not to be reported outside the scientific community? The issue here isn't fraud as such – which under "normal" circumstances is easily sniffed out by laypeople and experts alike – but the deviation from standard "truths" in a field of research by novel ideas and methods of inquiry. I doubt we are willing to pay the full price of conformity to avoid minimal fraud. Perhaps this is where the general public ought be educated to the fact that unlike the conclusive results it expects from scientific research (some outdated notion of certainty), the very nature of technoscientific research invites revisions and additions, and in this sense remains open-ended rather than definitive.

How prevalent is scientific fraud? Daniel Koshland, editor of *Science*, wrote in 1987: "the cumulative nature of science means inevitable exposure usually in a rather short time... we must recognize that 99.9999 percent of reports are accurate and truthful, often in rapidly advancing frontiers where data are hard to collect." Judson rejoinder: "The world of science in Babbage's day was small, open, and close-knit. Perhaps *his* confidence had been justified. Koshland's assertion was recognized at once as egregiously foolish, for it can have no basis in fact, either way" (Ibid. 132; italics in the original). But then Grant correctly suggests that the upsurge in fraud is due to the upsurge in scientific research: "the increase in fraud cases implies nothing one way or the other about (a) whether the percentage of fraudulent scientists is rising or (b) whether science is slowly becoming better at policing itself so that more frauds are being caught" (2007, 29). The more activity, the more likely we are to find some problems associate with such an activity.

DOI: 10.1057/9781137519429.0005

Admitting that "The borderlines between fraud, self-deception, gullible acceptance of the fake, and the ideological corruption of science can be very blurred" (Ibid. 12), Grant still argues that we must address this issue openly and in public forums, following the democratic ethos of science. These terms are of course conceptually problematic and deserve elaboration. Outright personal or institutional fraud is much more morally objectionable than self-deception, even if the results look alike. Likewise, gullible acceptance of the fake comes closer to self-deception, but in this case, such acceptance relies on a communal apparatus and the credibility of colleagues and professional organizations. Moreover, the ideological corruption of science is more broadly understood in the changing expectations of scientists and the compromises they are willing to make under the changed conditions of a neoliberal democratic state that measures its commitment to science in monetary terms. Though Greenberg agrees with this, he isn't as worried as Grant with the ongoing corruption in science. Noting the role of the US Office of Research Integrity, "which is specifically concerned with a narrowly defined band of offenses: fabrication, falsification, or plagiarism in proposing, performing, or reviewing research, or in reporting research results," he reports the declining numbers of such offenses (2007, 260–264). As we continue to think about the compromises scientists might be making willingly or unwillingly in order to survive and flourish in their respective areas of research, we should keep a vigilant eye on how pervasive fraud has become among the participants of the scientific enterprise.

Perhaps all critics of scientific fraud start with echoes of Judson's insistence on the centrality of truth in science, the kind of truth already invoked by Shapin's gentlemen of science, and one that provides trust among scientists and the general public. Perhaps when certitude is replaced with the expansion of knowledge, and unanimity with autonomy and democracy, as Agassi advocates, we can refocus on ways by which the scientific endeavor can be seen from a broad historical perspective, one where mistakes are made explicit and corrected, and where fraud is eventually uncovered and discarded. Trust among scientists should be institutionalized (regardless of the competitive nature of third-party funding of research) and become the foundation of the relationship between and interaction among members of the scientific community and the public at large. With this in mind, government regulatory agencies will simply have to provide oversight services, ensuring

DOI: 10.1057/9781137519429.0005

accountability and transparency rather than worrying about the veracity of what science accomplishes.

The compromises forced on the scientific community in the age of Big Science in the name of patriotism or nationalism spilled over, as mentioned earlier, to a marketplace culture of neoliberal ideology that endorsed such large projects. When the spillover consumed economically driven Big Science projects, partially funded by government agencies and partially funded by large corporations, questions of regulation and oversight have continued to puzzle the public: in whose name should we demand transparency and accountability? What compromises to national security might occur when public scrutiny is expected? But any public pressure for greater accountability and transparency will not only undermine the potential for fraud and misrepresentation of scientific research, but may also contribute to rehabilitate the idealistic expectations of scientists as angels, as pursuing a calling that follows the ethos of science on behalf of humanity. The call for (angelic) integrity might even work as an incentive for other professionals to rethink the financial boundaries and moral principles to which they should adhere. Taking required courses in business or engineering ethics does little to infuse a culture of integrity in students who are already pressured at the university to conform to the expectations of their laboratory taskmasters (or teachers); one can only imagine in horror how irrelevant the content of these courses becomes once they join the technoscientific workforce. With this in mind, we can now proceed to the next chapter, where corporate Big Science is discussed.

DOI: 10.1057/9781137519429.0005

3
Big Money: Setting Research Agendas

Abstract: *This chapter argues that just as national interests and security concerns dominate Big Science, so do financial motives and procedures of the marketplace control the research agendas of the scientific community. Economic pressures can distort the ethos of science and direct scientists in universities as well as industrial complexes into areas of research that they wouldn't have chosen on their own; moreover, these scientists might be forced to practice their trade in compromising ways, from keeping trade secrets to furbishing misleading data.*

Sassower, Raphael, *Compromising the Ideals of Science*, Basingstoke: Palgrave Macmillan, 2015. DOI: 10.1057/9781137519429.0006.

DOI: 10.1057/9781137519429.0006

The second kind of compromises and changing expectations of scientists happens regularly when researchers cater their work toward the goals of funding institutions, may they be private or public. This kind differs from the first kind discussed in the previous chapter because it has no patriotic or national security excuse, and it is completely determined by monetary incentives. Instead of participating in a patriotic activity for the common good, this kind of scientific compromise suggests that academic freedom, for example, is subordinated both in its direction and operations to the financial pressures of industry. The industrial–military–university complex overshadows any of the ideals of the scientific ethos. Unlike Big Science projects that come once every few decades, this kind (and its variants) is continuous; instead of national interests (as in the first kind), funding constraints determine research interests (without national pretexts in the second kind). The loss of intellectual autonomy and the concern with intellectual property rights color the conduct of technoscientists and the laboratories they inhabit.

To be driven by market expectations may not be a bad thing, as this way of thinking responds to the needs and wants of the public (however misguided they appear at times). But, as will become clear in due course, large projects may not be feasible for small investors or entrepreneurs, nor will they attract large corporations unless great profits are in store. As will be mentioned in Chapter 5, there might be unusual attempts by billionaires to discover new methods of research and development with a minimal profit motive, but this is an anomaly outside the established corporate structure. We can only wonder in this context whether or not the relative success of crowd funding (which eschews the authoritative and oppressive trappings of venture capitalists) may find its way to underwriting basic scientific research or be limited to the development of gadgets that have immediate appeal to those donating to the proposed projects (Sassower 2013b, Ch. 3, III).

The political economy of science

Over the past three centuries, various treatises on the functioning of the economy offered models according to which the state (and its various functions, including scientific research) should be organized. Beginning with Adam Smith, political economy was understood to incorporate moral and social elements that would set in place political and legal frameworks within which the marketplace could be organized. Beginning with *The Theory of Moral Sentiments* (1759), Smith embraced and espoused a vision of a village-sized community within which individuals' self-interest would naturally lead to the increase in the well-being of all others. Finessing the question of altruism, Smith cleverly described how a community would enhance the moral sentiments of its members, being watched by an *Impartial Spectator*. Is this "spectator" an image of God or one's conscience (or the Freudian super-ego)? Either way, for Smith it was clear that under such conditions – ongoing inter-action among neighbors watching each other and being watched by an impartial power that constantly judges their activities – social harmony could be achieved. Is this a description of how people in fact interact, or rather a prescription of how they should behave?

As a blueprint, Smith's early work suggests a social and moral frame-work according to which his later work, *The Nature and Causes of the Wealth of Nations* (1776), makes perfect sense. If we are already behaving morally, being socially responsible for our own actions and those of our neighbors, then there is no reason the *Invisible Hand* shouldn't suffice to guide our market exchanges. In other words, the Enlightenment ideals of individual freedom and equality are easily incorporated into a market-place where everyone is able and willing to interact (on equal terms and voluntarily) with everyone else. Every member of the community is free to enter or exit the market without prejudice or penalty; and when there, each enjoys the efficiency of the division of labor, abiding by the fluctua-tion of prices according to the supply and demand of goods and services. An ideal model is thereby introduced, economic as well as political, social, and moral.

Political economy is as much a way of thinking as it has been associ-ated with a period in economic history when economic variables were understood to be intimately related to their ideological background. In the 19th century, Karl Marx advocated his ideas in this spirit: a seismic political shift (revolution) must take place in order to provide a more equitable and less exploitive and alienating society. His critique of clas-sical capitalism is as much an economic critique about the inherent inadequacies of market capitalism and its inevitable failure, as a social and moral critique of the concentration of wealth and income in the hands of the few (see a contemporary discussion of some of the nature of wealth inequality in Piketty 2014). For him, just as for Smith before him, economic relations are supposed to be morally justifiable and

DOI: 10.1057/9781137519429.0006

establish social coherence and political stability. When either model fails to adhere to these principles, the entire system collapses. Marx's socialist (and communist) ideals were not simply a critical response to capitalist abuses, but actually the utopian starting point from which to devise alternative models for human cohabitation and interaction.

When thinking about science within this framework, we should right away switch to the more appropriate term *technoscience* (as defined previously), as this term captures more accurately the blurring of the distinctions between science and technology. On this view science no longer retains its alleged pure and theoretical designation in comparison to technology's applied status; instead of these untenable demarcations, they inform each other and are seen seamlessly continuing from one phase to another, and then back again. So, how should we appreciate the role of technoscience in political economy? There are those who, following the linear model discussed earlier, argue that technoscience plays an integral role in the economy and has become the driving force for its progress and growth. They cite the cases that begin with the Industrial Revolution and continue to the present with the Digital Age. Every new developmental phase of technoscience is associated with or can be directly traced to improvements in the economic conditions of the state; technoscientific prowess has become the yardstick against which developed countries measure their status in comparison to those less-developed ones that fall behind them (in terms of annual income per capita, for example). The level of government support varies from one country to the next, but most developed countries appreciate the economic benefits to growth and sustainability that technoscience provides, from ideas and ways of thinking about the environment to health care, transportation, communication, and entertainment. In fact, the very distinction between the developed and less-developed or developing countries hinges to a large extent on their respective reliance on and use of technoscientific innovations.

Practical considerations of the benefits of technoscience express the ideology of the role of the state to further the economic welfare of its citizens, regardless of any nationalistic contours it might belie as well. Ideology informs government funding agencies insofar as certain priorities are foregrounded and others neglected (more for the military and less for parks, or more for national monuments and less for libraries). In this context, it seems that there is great appreciation for the central function of universities to train future generations of scientists and provide

DOI: 10.1057/9781137519429.0006

solutions to pressing social problems. But these economic considerations, though assessed in national and functional terms, are in fact manifested through the corporate world of super- or hyper-capitalism, so that there emerges a divergence between the interests of the state and those of industry. What was once a natural alliance between the two – what's good for General Motors is good for America, as the saying went half a century ago in the US – from subsidies to regulations, has fallen apart. Instead, we see ongoing battles that pit national interests – workplace safety, health-care provision, clean air – against the pursuit of profits at all costs (where cutting corners and polluting are commonplace). If Smith envisioned a division of labor that increases productivity, late capitalism envisions a division of responsibility that burdens the state (bailout subsidies) while freeing industry from any responsibility to its workers and customers alike. In this climate, the activities of technoscience, from research and development to engineered implementations, must be understood in a political economy rather than national framework.

If there is a divergence of interests between the state and industry when it comes to technoscience, it becomes obvious how the expectations of scientists change along these lines as well. Some scientists will be lured to undertake industrial projects that promise rich rewards and participate knowingly or inadvertently in problematic and even corrupting or publicly damaging activities regardless of state mandates and regulatory constraints. As political economy has transformed into economics and econometrics, as the consideration of the ideological underpinnings of social organizations are replaced with a narrow and quantifiable measure of efficiency and success, moral principles are inadvertently compromised. This is the climate that continues to dominate Western democracies since World War II (perhaps with some exceptions in some European countries). It is also a cultural climate that replaces patriotic self-sacrifice with the protection of intellectual property, for example, for the sake of profit maximization.

Science at the university

As early as 1918, Thorstein Veblen decried the commercialization of the university. For him this was an outrageous development that eviscerated the sanctuary-like qualities accompanying both esoteric musings and the custodial responsibilities of academics in their ivory towers. The

DOI: 10.1057/9781137519429.0006

development of the atomic bomb included various academic institutions, but not until the admission that the university was at the disposal of industrial needs (rather than those of the state) has the academy indeed become in Clark Kerr's sense a "multiversity" (1963). Perhaps this was an admission that the interests of state capitalism were fully aligned with those of the markets; perhaps it was a foray into a new commercialized state of affairs where economic thinking overshadowed all others. Either way, the university was incorporated into the markets, as Sheila Slaughter and Larry Leslie claim, and therefore turned into "academic capitalism" (1997, 4–8). This means that capitalism – with its attendant principles of market efficiencies due to division of labor, price determination in light of supply and demand allocations, and profit maximization – engulfs every facet of the state, including its knowledge-production institutions. What happened to the scientific commons? The scientific commons was defined as the body of knowledge that is open to all scientists and citizens alike, and this view was being threatened by the onslaught of intellectual property rights (patents and copyrights) so that the scientific ethos, with its claims to universality and communalism, was bound to come under attack.

To be clear, a great deal has been written about the British notion of *Commons* and its application to contemporary digital technologies, for example, or the rethinking of economic relations in the 21st century. The notion or ideal of the commons has been understood as one of the oldest forms of self-organizing governance, and one of the most promising futuristic ways of thinking about a "collaborative commons," as Jeremy Rifkin (2014), for one, conceives of it. From the Free Software Movement of the 1980s whose agenda was to allow free and open source for compu-ter code, shared by anyone for whatever reason so long as no restrictions and profits were exerted at the expense of others interested in using the same code, all the way to the Free Culture Movement of the late 20th and early 21st centuries, futurists have argued about the need to think in collaborative rather than individualistic terms. Their ideals may seem less radical (or communist in an orthodox Marxist sense) when considering the millennia generation's attitudes toward access to information and entertainment rather than an outright private ownership of its media. The point here is to illustrate how technoscientific developments may shake the foundation of the ideology of market capitalism, but in doing so may skip the re-envisioning of the function and structure of academic science. Back to the academy and its practitioners (Press and Washburn 2000).

DOI: 10.1057/9781137519429.0006

As Slaughter and Leslie continue to explain, the shift in attitudes and practices of academics wasn't simply "replacing altruism with a concern for profits. Rather, they elided altruism and profit, viewing profit making as a means to serve their unit, do science, and serve the common good" (Ibid. 179). This portrayal allows scientists as academics to retain the ethos of science while translating its application in this context quite easily. They aren't abdicating their responsibilities as angelic and idealistic scientists, but instead harnessing the "value" of technoscientific products into prescribed idealized goals (ones they can endorse). These are the changing expectations of scientists: can they remain true to their calling while being pressured to provide ready-to-use items for the marketplace? This question is of course complicated by the fact that as we have already shown, by the late 20th century the distinction between applied science/technology (the focus of industry) and basic/pure science (under the custodial guardianship of the academy) has become blurred in the entrepreneurial university system. The provision of research that could be perceived "public," in the American sense of the term or the "commons," as the British term denotes, is in full compliance with and at the service of the markets. There is no clear separation for three reasons: first, government funding demands private licensing; second, academic research includes public and private funding; and third, grants determine research direction. Under such conditions, public interests are supposed to be perfectly aligned with private ones and vice versa, and commercial applications are seen as almost intrinsic to the process of inquiry.

Elizabeth Popp Berman reminds us that the "economization" of technoscientific research and development not only changes the language used in describing the activities of technoscience, it also changes the cultural setting within which government policies are administered and the legal frameworks that ensure their dissemination (2014). This means that if the state is considered in economic terms, and if the university is part of its functional apparatus, then it stands to reason that the academy is understood in economic rather than intellectual terms. In this sense, then, the changing expectation of scientists is less of a personal, deliberate complicity of the scientific community with the temptations of capitalist interests. Instead, it's a structural setting of the university system, already cited earlier in the 20th century by Veblen and Weber, which bows to economic pressures and cannot remain detached or insulated from the overwhelming demands of market capitalism.

DOI: 10.1057/9781137519429.0006

But is the university really benefitting financially from economizing its agenda? As Mirowski persuasively argues, "*very few universities make any money whatsoever, much less serious revenue, from their management of their IP assets.* This curious situation forces us to contemplate the notion that the modern insistence upon the commercialization of science is more likely to have ideological, rather than simple behavioral, motivations" (2011, 182; italics in the original). If the university system as a whole and individual universities within it do not profit from this commercial feature of their restructuring, then the financial incentive isn't a factor at all. This illustrates that the very assessment of the grounds that lure academic administration to bow to commercial interests may have less to do with actual revenue and profits and much more with their mindset, a mindset that has been accustomed to think in monetary rather than intellectual terms about the mission of the academy.

Though focusing on the transformations taking place in the biomedical sciences in the university system, Sheldon Krimsky clearly supports the views expressed earlier and is quite critical of the potential problems associated with the commercialization of the university system. There is a concern that the pressure of academic entrepreneurship will transform knowledge production into a commodity where expertise is bought and sold to the highest bidders. Substituting "protected enclaves" of "sources of enlightenment" with "instruments of wealth," the newly founded university–industry complex involves "collaborations, technology transfer, and intellectual property" (2003, 2). Collaboration in and of itself is neither corrupt nor necessarily entails corruption of academic techno-scientists; it could be the hallmark of the ethos of science, as described by Merton. But when the collaboration is limited to funder and the funded rather than the entire scientific community, then a corruption of an ideal is in plain view. Technology transfer, though, could be understood as the pressure to provide market-ready products, and as such pressures technoscientists to conform to the dictates of their funding masters, rather than consider broader potential for scientific research. And finally, intellectual property rights subvert the very notion of research as a public good or commons into the entrepreneurial domain, where secrecy dominates all scientific communication, thereby clearly undermining the ethos of science. Under these conditions, technoscientists are bound to make compromises: what is expected of them from those in control of their livelihood may undermine the principle they wish to abide by. Krimsky represents a modern restatement of Veblen's and Weber's views

DOI: 10.1057/9781137519429.0006

about the very nature of the university and its social role within the state. If academic scientists are encouraged to pursue technology transfer and establish private companies that can exploit the fruits of their research, then obviously the character of the university has radically changed.

As we shall see more starkly in the next chapter, once commercial interests overwhelm academic technoscience, questions of conflict of interest, for example, are managed rather than avoided, prevented, or eliminated. Once academics buy into the economic ethos instead of their idealized scientific ethos, the temptation to compromise is great. And legal sanctions, when they appear, are both at the very margins of their practices and often treated as negligible nuisance. Krimsky provides guidelines to mitigate the problems associated with conflicts of interest at the various levels and points of contact between scientists and their sponsors, and emphasizes as a symptomatic illustration the Supreme Court case *Diamond v. Chakrabarty* (1980), which ruled that "genetically modified bacteria were patentable in-and-of-themselves, apart from the process in which they were used" (Ibid. 30). If courts of law bow to market pressures, if they are willing to grant patent protection when none was imagined only decades ago, how can academics resist these and future transfer pressures?

Even the Office of Technology Assessment prophetically predicted in 1988 that: "It is possible that the university–industry relationships could adversely affect the academic environment of universities by inhibiting free exchange of scientific information, undermining interdepartmental cooperation, creating conflict among peers, or delaying or impeding publication of research results. Furthermore, directed funding could indirectly affect the type of basic research done in universities, decreasing university scientists' interests in basic studies with no commercial payoff" (Ibid. 31). Against this view, we find Greenberg, for example, who argues that the "squawking" [of critics of the academic–business enterprise] is disproportionate to the reality … There's no truth, however, to the frequent, wholesale depiction of university-based science as a passive appendage of corporate America" (2007, 43–45). Faced with these two opposing assessments of the situation, we may want to take a second look. It may be that the alarm bells should be sounded, but more as a warning than an indictment. Likewise, it may be that both sides are disagreeing on the *degree* to which this is a problem rather than there is a problem at all. Perhaps some part of university-based research is indeed completely beholden to its funding sources, but not all. Perhaps

DOI: 10.1057/9781137519429.0006

university researchers have figured ways in which they can serve the interests of their sponsors and underwriters while still maintaining their integrity and their own research agendas. My guess is that the truth lies somewhere in between those critics who condemn all of university science as being commercialized and those defenders of the faith, so to speak, who retain their faith that technoscientists haven't lost their soul, haven't all become fallen angels.

Whether remaining more critical or more sanguine about the regulatory and legal obstacles to a wholesale transformation of academic research into the commercial domain, I'm still concerned with the internal mechanisms that are set by academic institutions to control their members' potential corruption. There are no national standards adopted by university administrators; they remain discretionary and *ad hoc*. Is this an institutional failure? Is there a way to reverse the commercial tide? Or, are we dealing here with a public relation campaign (in Agassi's sense) to ensure the continued flow of money to the academy? The ability to manage conflicts of interest appears to be, at best, superficial. "Shame and embarrassment exercise great force in academic and scientific affairs. Pride plays a big role, too. Scientists, their managers, and their institutions normally care deeply about their reputations" (Krimsky 2007, 258). This is the case because on some level we expect the technoscientific community to behave better (socially and morally) than the society in which it works.

If moral sentiments are relevant, as Smith already argued, then we can imagine that some modes of behavior within the sacred halls of the academy warrant scrutiny. Have we gone too far in the direction of Berman's economization of the academy? Are we at a point of no return, especially when state support for the knowledge industry is waning under recession-like conditions? Incidentally, even though the economic conditions in the US have slowly but surely improved in the US in the past five years, local support of state universities remains at its historical lows. (For example, the state of Colorado provides around 5% of the budget of its flagship University of Colorado system; the rest comes from tuition and sponsored research.) Whereas for some, reputation might ring alarm bells about the "unholy" alliance between the university and capitalist markets, for others, this may signal a change of heart: "they produce strong, perhaps irresistible, academic insistence on shared governance over use of industrial money; quick, if not immediate, publication of the results; and adherence to academe's concept of the rules

DOI: 10.1057/9781137519429.0006

of the game" (Ibid. 48). What may seem like a benign acceptance of the realities of external funding eventually turns into routine, institutionalized recognition that some modicum of academic integrity is worthy of public perception.

This general indictment of the university system (with potential for improvement) spills into American politics, as James Savage reports. Earmark appropriations by Congress – added amendments to legislation that serve as inducements for garnering majority votes – is a common way to avoid the peer-review process either because it is dysfunctional – rewarding a small elite no matter the merits – or as a way to avoid the rigors of peer review altogether. An earmark, he explains, is "a legislative provision that designates special consideration, treatment, funding, or rules for federal agencies or beneficiaries" (1999, 6). Eschewing peer-review panels that would certify the credibility of research and its needed infrastructure in universities, academic earmarking may be either a system that skirts the standards of the academy (in terms of rewards for merit) or one that corrects the biases built into the academic hierarchy both within and among universities (where better endowed and more famous universities always win grants while the lesser known ones always lose). Of course not everyone is alarmed by earmarked funding because no proof has been provided that there is a direct correlation between political appropriation and fraudulent scientific research results.

Appropriating falsely Smith's notion of self-interest as the guide for social welfare, neoliberal ideologues misapply an economic model where individualism is paramount (of persons and corporations) laying to waste the necessary political conditions that allow the marketplace to strive (except when it gets to protecting intellectual property). Even when the language of neoliberalism is softened, as Berman (2014) eloquently undertakes to demonstrate, the central role of government to administer (neo-Marxist planning and coordination) policies that ensure certain outcomes remains intact. The ideological thrust of entrepreneurship and the efficiency of markets remain undeniable hallmarks so that economic rationality is deployed in policy debates. "Economic rationalization," as Berman calls it, is the standard yardstick used by policymakers and the administrators of academic science (Berman 2012). Is this way of approaching technoscience undertaken on behalf of improving human welfare? Is technoscientific "input" measurable? Even when it is, what impact does it have on "outputs" (assuming they are clearly defined)? It is no wonder that the university system, the wedge between the state and

DOI: 10.1057/9781137519429.0006

the economy or the engine that moves the entire economic train, has taken such a significant role in this debate; likewise, it is no wonder that the expectations of scientists have changed with the changing character of the academy. Academic technoscientists are swept in the frenzy of job-training and increased external funding to sustain academic bureaucracies and appease those concerned with tuition hikes and post-academic job searches. Some of them are victims of the system, some are willing participants, whereas still others are the ones initiating and promoting this compromised and compromising system.

Though most concerns with scientists falling from their angelic grace, so to speak, at the university level, are associated with academic capitalism, Savage's critique alerts us to another problem. Political power brokers are willing to overlook whatever the ethos of science recommends in the name of expediency (and are thus also motivated to offer congressional earmarks). Is expediency the only proper way to allocate state funds to worthy institutions? No matter his own (neoliberal) ideological biases, Savage is right on the mark in the assessment that in avoiding proper scrutiny, earmarked university funding is both morally and practically flawed. He concludes by arguing that if earmarking is indeed not the most efficient allocation process of funding, then of course the nation suffers from it. In this sense, then, it's not academic technoscientists who are at fault at all, but rather a flawed system of the allocation of state funds.

Case studies of industrial–scientific corruption

The focus on cases where flagrant deceit has overtaken scientific research and development can be divided into five areas: tobacco, asbestos, biotechnology and pharmaceutical, food and agriculture, and oil and gas (though other classifications might do just as well, as seen in Ashton and Laura 1998). In all of these cases, critics note the kind of "sponsorship-bias" where generated results are suited to the expectation of those who fund them, as well as the suppression of inconvenient results that seem unfavorable to or undermine expected outcomes (Parkinson and Langley 2009). This sentiment is an echo of two previously stated concerns with the commercialization of science in and outside the academy (Broad and Wade 1982). There were those who after World War II were concerned about the fact that more than half of the scientists and engineers in the US were on the

DOI: 10.1057/9781137519429.0006

military payroll, in so many words, working on warfare projects (Singh and Gomatam 1987). Then there were the sociologists of science who were worried about the fabrication of scientific facts that has become routine in laboratory life (Latour and Woolgar 1979). What they all have in common is the concern with money becoming the driving force of the scientific enterprise, so that it transforms the expectations of scientists themselves and the public that supports them.

As Rampton and Stauber argue in the contemporary use of expertise, when we speak of the scientific enterprise we should be clear that we are dealing with only a sub-set of scientists, those who belong to a "particular class of experts who specialize in the management of perception itself – namely, the public relation industry" (2001, 2). What is the issue here, as Agassi already noted, is public perception and regulatory compliance, rather than any debate over the merits of science itself. What is bothersome is the prejudice that underlies the use of experts in public relations campaigns, namely, that the public is easily swayed by emotional appeals and would not be able to critically evaluate scientific claims. Whether this is empirically true or not (see Kahneman 2012) is less relevant than the fact that it is the prevailing assumption according to which the public relations of science is pitched by that particular class of scientists. At times, this manifests itself, as we mentioned earlier, in popular media depictions of the idealized images of scientists and their angelic contributions to humanity's store of knowledge. At other times, this means, for example, that the biotech industry can be viewed in two ways: "In one case, we have tremendous public support – we can be viewed as farmers bringing new varieties and improved foods to consumers. But if we do not position ourselves and our products correctly, we can just as easily be viewed in the same class as Hitler and Frankenstein" (Rampton and Stauber 2001, 54). Angels or fallen angels, saviors or villains, technoscientists can enjoy or suffer from public perceptions, depending on who their promoters or critics happen to be.

The shift from concerns over the truth to how to manipulate data to elicit the most favorable (emotional) public response is coupled with risk assessment. Cost–benefit analysis is common in economic models; it has been used in the business world to direct and control decision-making processes. Once all risks and benefits are quantified, they can be measured against each other (classic technique of double-entry bookkeeping) to evaluate the "bottom line" loss or profit of a project. No different from 19th-century utilitarian theory (Jeremy Bentham (1748–1832) and

DOI: 10.1057/9781137519429.0006

James Mill (1773–1836) come to mind here), this way of thinking sounds reasonable enough for technoscience overall and especially when policies are chosen and implemented. But as Rampton and Stauber remind us, it's not only a question of when "safe" is safe enough; what is forgotten at times when these calculations are made is that it's not the same group which benefits in relation to the one taking the risk: "Risk assessment, it is now clear, promises what it cannot deliver, and so is misleading at best and fraudulent at worst" (Ibid. 107–109). In other words, those benefitting might be only those who own the industries associated with the project assessment, whereas those suffering are the residents at their operating locations (and at times even those living far away, namely, residents suffering pollution or consumers of the products in faraway places). Questions of truth and certainty, safety and risk assessment, plague every facet of technoscience, even though in each case there are unique features that highlight the manners in which scientific corruption is evident: in some cases it's clear and obvious – this plant dumped this toxin into this river; in others, the chain of production, distribution, and consumption is too long to be able to ascertain which link in the chain is problematic or toxic, which particular stage caused a particular effect.

The next chapter will be devoted to the biotechnology industry, including Big Pharmaceutical companies, so here I skip this enormous moral minefield of technoscientific research and development. As I hope to demonstrate later, all areas where changes in expectations are taking place and pose grave problems are related to health care, in the broadest sense of the term. So, when covering fraud, deceit, and outright corruption of the ideals of science, we find that asbestos and tobacco bring about health hazards in their consumption just as agricultural and energy concerns eventually have health impacts on people and their environment. The reason to separate the pharmaceutical industry from all other industries is because it illustrates a more pernicious sense of scientists falling from grace, one unfathomed some 50 years ago.

Asbestos

Though each of these cases deserves a lengthy survey, in what follows I briefly focus on the deceptive elements perpetrated by scientists rather than captains of industry or their lobbying agents. In the case of asbestos, it's easier now to reflect on how some scientists were complicit in hiding the empirical evidence linked to the health hazards of asbestos, because

DOI: 10.1057/9781137519429.0006

the story is well known and legal measures are in place to protect the public from this hazard. Michael Bowker begins his devastating report on asbestos by reminding us that "although medical evidence long ago proved conclusively that asbestos is lethal, it remains a legal ingredient in more than three thousand products nationwide" (2003, 15). From the start, there are some oddities when we deal with the regulations covering asbestos: "The EPA, OSHA, and other federal and state regulatory agencies each have their own rules regarding the levels of asbestos fibers that are allowable and considered safe in homes and in the workplace. At the same time, the EPA's Web site states: 'there is no known safe exposure level of asbestos'" (Ibid. 17). From the start, we find a logical (hence scientific) contradiction, if not a confusion: either no level of safety for asbestos exists, and therefore we should outlaw its use completely; or, there is an acceptable level, and we should specify its boundary conditions. Which is it? Unlike the US, it should be noted that the UK enacted laws to regulate asbestos already in 1931 and 1932 (Ibid. 48).

The tale of woes chronicled by Bowker dates back to the 1930s: "Asbestos-related diseases have virtually been ignored by the government, and little federal money has been spent on research." The reason for this has to do with the "corporate, legal, and political web of deceit and duplicity" (Ibid. 21). It's interesting to note that the simple empirical fact of miners becoming sick wasn't sufficient to sound the alarm or induce more research. Instead, "when miners and workers who smoked came down with lung cancer, the disease was always written off to the tobacco use. Many of the asbestos companies underwent fierce campaigns to reduce tobacco use in their plants, even as they did nothing to curb asbestos exposure" (Ibid. 39). So, one mode of deception, underwritten as it were by scientists and medical researchers, had to do with deflecting the cause of cancer or other diseases from asbestos to tobacco. As Proctor concludes, "The history of asbestos is a history of scientific deception joining hands with industrial malfeasance" (1995, 122). This is different from just trying to embark on a doubt campaign that will stall regulatory action, or from accepting funding from a company whose eventual technology remain a proprietary secret.

The other mode of deception, one more directly requiring the explicit accommodation of scientists, had to do with research funding. Sanarac Labs were funded by industry, and prevented publication of results that showed asbestos exposure dangers in 1937 (Bowker 2003, 53). This ongoing deception and corruption of science could be publicly observed in the annual symposia and conferences (especially the 1952 one in New York City) where the collusion of industry, insurance companies, and government regulation failed to warn against asbestos exposure (Ibid. 93–94). Was this collusion still beholden to the anti-communist ideology, which, after World War II, linked any government regulation to the horrors of fascist tyranny? Or, was it instead, simply a market-driven means by which to silence any alarm bell and subdue any need for closing down an entire industry? Who was to stop this capitalist engine of progress and prosperity? Were all technoscientists capable, both methodologically and financially, to stand up to their corporate paymasters? Was their idealized ethos a worthy public relation ploy amidst the hustle and bustle of commercialism, a form of deflection rather than responsibility and integrity?

Tobacco

The second set of case studies deals with the tobacco industry, a major economic force for most of the history of the US. As the 20th century came to a close, according to Proctor, the entire remediation process of hazardous materials has taken on a life of its own. Large industries have emerged to abate and remediate environmental hazards such as asbestos, lead, and radon. Though this means that the focus on cleaning up past mistakes is in full swing, "the commercialization of environmental troubles has led to novel conflicts of interest, along with unprecedented stakes in the exaggeration or diminution (often by fiat) of environmental hazards." Experts are invited to make decisions, as we have seen earlier, and when doing so, they disagree not only about the gravity of the hazard, but also how it should be handled, if at all.

Instead of highlighting the public relations dimension of science (as Agassi, Rampton, and Stauber do), Proctor alerts us to the "social construction of ignorance" in the name of science as a way to avoid dealing with real scientific issues (Ibid. 8–9), and the importance of trade associations as fostering "science as advertisement" or public relations campaigns. Trade associations have become adept at shifting knowledge into ignorance insofar as ignorance can be managed so as to cast doubt on any expert claim. The reason Proctor is so concerned with and critical of trade associations is that they exemplify how scientific work can be subverted: "trade associations are able to combine research and advocacy in ways that insulate members from charges of individual bias or legal liability" (Ibid. 102–104). Members here include tobacco companies

DOI: 10.1057/9781137519429.0006

as well as their scientific teams, whether on the payroll in-house or at universities and corporate-funded research centers and institutes.

Though there are various industries that are responsible for the increase in cancer cases, Proctor pays greater attention to tobacco because its manufacturers have been pioneers in managing uncertainty (not to mention slavery in earlier times). This kind of management is not limited to the ranks of public relations firms that present data in ways that manipulate public perception, but also permeates the ranks of a particular class of scientists, as Rampton and Stauber called them. For them the idea of "science as propaganda" is a way to demand "balance" in alleged scientific debates over the hazard of smoking or the causes of cancer. Once a balanced debate is established in the name of debunking so-called junk science, there is an easy trajectory, over time, to sow doubt in the minds of the public as to the real causes of cancer, and infuse enough doubt to undermine any meaningful scientific claim. This is important in relation to the goal of science which is supposedly the discovery and testing of putative truth claims after a great deal of research, rather than to maintain an ongoing debate with balanced arguments (reasonable or not) in the name of courtroom fairness and justice. In other words, the appeal of science in general and the scientific method in particular is its basis in the empirical world, in the measurable and testable collection of facts and experiments, unlike the interminable debates of ideologues who change the terms of the debate whenever they feel at a loss or about to be defeated.

As mentioned already earlier in tactical terms, Proctor enumerates various rhetorical strategies used by private corporations to avoid regulation or admission that they are culpable for any public hazards. The *first* has to do with the notion of uncertainty where the evidence is claimed to be "ambiguous, inconclusive, or incomplete" which necessitates more exploration and study. The *second* has to do with inaction because more study stalls regulatory action. The *third* has to do with the political climate of deregulation (from the Reagan Administration to the Bush and Clinton Administrations that followed, and Thatcher's in the UK). When more scientific research is called for, more time can elapse before implementing regulatory standards, or what is also called "paralysis by analysis." The *fourth* has to do with the claim that the company is making progress, no matter how slow. The *fifth* has to do with the mantle of scientific respectability taken on by trade associations as being neutral, balanced, and levelheaded. "The continual reference to science (the

DOI: 10.1057/9781137519429.0006

words *science* or *scientific* appear at least once in any paragraph of many AIHC [American Industrial Health Council] publications) gives the council an aura of respectable neutrality, an image of standing above the petty interests of profit and personal gain" (Ibid. 125–131).

After years of federal litigation and after billions of dollars have been set aside by the tobacco industry, cigarettes are still sold and consumed in the US and around the world. Despite scientific evidence about addiction and the effects of second-hand smoke, there is sufficient legal latitude to allow commercial interests to dictate the conditions under which to prohibit smoking. As Proctor explains, questions of health are turned into questions of free speech. But is free speech in the Constitutional sense really at stake here? We commonly appreciate the defense of free speech in terms of those minority voices who may propose unfashionable ideas; in the debates of the scientific credibility of this or that position concerning the effects of tobacco consumption, the contention that free speech is impeded seems unwarranted because hardly anyone is silencing the scientific voices paid for by industry or the work of advertisers. On the contrary, it is they who dominate the debate and silence or marginalize their critics.

Likewise, "a misplaced call for precision can also be wielded as a political tool – as when calls are made to delay regulation until endlessly more elaborate studies confirm that a substance is hazardous." This means that notions of uncertainty are overly exploited, that the need for verification is exaggerated, so that the legitimate concern for certainty and the decisive data that would settle any doubts are lost in the process. Doubt and uncertainty take on a legal and social function in warding off corporate challenges and potential changes, rather than being scientifically understood as the first steps toward greater certainty and the elimination of doubt (Ibid.). This seems to be a blatant instance of how the ideals of science have been subverted and what compromises scientists have had to make as expectations of their ability to provide credible cover for their paymasters. Admittedly, this doesn't mean the subversion of the entire scientific edifice and the compromises of the entire scientific community; but when such behavior is publicly displayed (and covered as scandals in the popular media), the trust in science and its purported ideals are shaken.

Food

The third set of case studies deals with agriculture and the food industry. Marie-Monique Robin titled her sensational book *The World According to*

DOI: 10.1057/9781137519429.0006

Monsanto (2010) to highlight the fact that this single corporate giant isn't simply providing goods and services, but is in fact framing the entire global agricultural industry discourse in its own image. What becomes evident is that, with full protection of the American legal system, one company can genetically modify seeds, sell them to farmers and farming entities, and control all the distribution and production of crops. Under these peculiar conditions (completely outside of the classical economic model), nature itself is being subjugated to a monopoly of corporate interests. Natural phenomena, such as seed regeneration, are stopped so that farmers must buy new seeds every season. Moreover, with monopoly power over the production of foodstuff, companies like Monsanto have lobbying power over federal Farm Bills that provide subsidies to particular segments of the agricultural industry as well as determine what counts as a healthy diet.

Parts of the scientific community bear direct responsibility to what happens, from farm policies that sanction particular commercial interests, all the way to the promotion of the nutritional value of what consumers eat. As Marion Nestle reports, there is a revolving door between experts who work for private companies one year and in the next take positions in the federal agencies that regulate them (2002, 99–100). For most Americans (and perhaps for most Europeans), the major sources of nutrition advice are media outlets and the advertisement of the food industry itself. Government agencies are either absent from direct involvement at all levels of public education – perhaps because of Cold War worries over totalitarian indoctrination – or leave it to private interests to suggest what is healthy and should become public policy (in school cafeterias, for example, and military bases).

Nestle argues that personal beliefs overshadow scientific data. Consumers are convinced that dietary supplements offer some sort of a nutritional insurance. This belief is cleverly framed as one's right to choose whatever one eats without government intervention. Whether or not these supplements are effective becomes a secondary question that isn't often asked (Biss 2014). Once again, as we saw with Proctor's analysis of the "cancer wars," scientific questions are translated into freedom of speech and action. "Science-based belief systems ... explain why federal and private health agencies rarely recommend nutritional supplements as a replacement for foods" (Ibid. 292). What should one do with the *1995 Dietary Guidelines* that advised consumers that "daily vitamin and mineral supplements ... are considered safe, but are usually not needed by

DOI: 10.1057/9781137519429.0006

people who eat the variety of foods depicted in the *Food Guide Pyramid*?" Is the "safety" issue foregrounded so as to overshadow the "need" factor? The most Nestle can admit is that science-based arguments have indeed shed some level of skepticism among professional dieticians, but have not deterred consumers from following the directions of supplement manufacturers.

The debate over supplements also includes another paradox, as Paul Offit and Sarah Erush remind us: although the Joint Commission which is responsible for hospital accreditation in the US "requires that dietary supplements be treated like drugs," the Food and Drug Administration "doesn't regulate dietary supplements as drugs" at all. They continue to list a variety of cases where outright fraud and danger were detected. In 2003, "researchers tested 'ayurvedic' remedies from health food stores throughout Boston" and found that "20 percent contained potentially harmful levels of lead, mercury or arsenic." In 2008, products were pulled off the shelves because "they were found to contain around 200 times more selenium (an element that some believe can help prevent cancer) than their labels said. People who ingested these products developed hair loss, muscle cramps, diarrhea, joint pain, fatigue and blisters." In 2013, "vitamins and minerals made by Purity First Health Products in Farmingdale, NY, were found to contain two powerful anabolic steroids. Some of the women who took them developed masculinizing symptoms like power voices and fewer menstrual periods." Such cases are so common that "the FDA estimates that approximately 50,000 adverse reactions to dietary supplements occur every year" (Offit and Erush 2013). Shouldn't we know about these cases and the dangers they pose to the public? Is it possible for the general consuming public to be sufficiently familiar with all of these data?

The food chain is long and cumbersome, as popularized by Eric Schlosser (2001), and as such becomes a formidable obstacle to those interested in learning about it. It includes what we feed cows and chickens, how we slaughter and process them (Sinclair 1905), all the way to the distribution chains of grocery stores and restaurants, fast and slow. It also incorporates genetically modified organisms (Nelson 2001, Smith 2007), the corporate power over public policy and food production, all the way to the way we think (under- or over-think) our relations with the food we consume. This isn't limited to the obsession over television food shows where chefs compete, but permeates the public-health debate over obesity and its causes. Health-care issues are finally understood in

DOI: 10.1057/9781137519429.0006

nutritional terms as much as genetic and environmental ones, so as to highlight the areas where direct preventive intervention can be rational and useful. Isn't this what the scientific ethos is all about?

Energy

The fourth area of case studies that illustrates the changed expectations and practices of scientists is linked to the oil and gas industry, and ranges from the fuel efficiency of the automobile industry all the way to energy independence, alternative fuel sources, and the debates over climate change and its impact on the environment. Just as the three previous areas are too broad to handle in brief, all that I offer here is an overview of the complicity and outright deception perpetuated by some scientists and engineers at the service of large corporations. One need not be a conspiracy theorist to believe that over the past 100 years more efficient combustion engines haven't been invented or even tried. Why have they not made it to the car market? What collusion between automobile and oil companies and their lobbyists has not been fully revealed? One can only imagine the legislative cover politicians have been willing to offer their sponsors over the past century. Why has fossil energy dominated world markets? Why have alternative energy sources been so underfunded and underutilized (Ashton & Laura 1998)? Except for a few countries around the world that strive for energy independence, including the US and Canada more recently, most developed countries are still struggling to achieve this status.

The contemporary quagmire of global warming and climate change requires its own book. Suffice here to say that scientists on both "sides" of the debate provide evidence they claim to be irrefutable and substantial enough to warrant the "other side" to change its mind if not its rhetoric (Irwin 1995, Ch. 6). It also suffices for us to observe recent climate trends and catastrophes to appreciate that something in fact is going on, regardless of how the data are analyzed. Assuming that global warming and climate change are "scientific" questions and not ideological ones (as they sometimes appear in the media), why can't we simply look at the facts and finalize our assessment or reach some level of agreement? Perhaps this is the most apparent and worldwide phenomenon that brings to a head the very essence of scientific credibility and its ethos. *If science fails to give us clear-cut answers, how does it differ from religion or ideology?* If the scientific enterprise is beholden to commercial forces larger than itself, if it remains the mouthpiece of those who pay to direct

DOI: 10.1057/9781137519429.0006

its pronouncements, then how does it differ from any public relation or advertisement enterprise? It could, of course, prove its mettle by arguing (even in business-like manner) that in order to hedge our bets we should pursue more than one track on the way to energy sustainability and independence. Or, it could insist that there are industrial answers that can be partially supported by scientific research, but also alternative ones. This kind of discourse would make much more sense than the rhetorical or ideological one, because silly claims could be readily falsified, and reasonable ones confirmed.

Obviously the scientific enterprise, as Rampton and Stauber claim, is at times in the service of the powers that be, and at other times propounding the ideals of democratic and communal association of knowledge seekers. This could definitely mean that "science had ceased to be merely a methodology and had become an ideology as well" (2001, 37). One can argue that this has been so all along, especially as science was set against the backdrop of a powerful church that under some circumstances demanded subservience to its dogmatic doctrine. Yet with the introduction of so-called *junk science* and the appropriation of the term in courts of law whenever an industry wants to discredit any of the opponents' research, what is happening in the 21st century is that standards of empirical testing are being ignored. To be sure, in the US, both in the popular media and in the courts, scientific data, research, or analysis that are considered to be spurious or fraudulent, suspect or fabricated, are called junk science. The pejorative connotation that is associated with this designation is highlighted when ideological interests come into play, such as with climate change debates, but in most cases the label isn't supported by claims about proper scientific standards of research. When these methods and standards of the scientific ethos are ignored, the well-established legitimacy of scientific methodology as a whole is being ignored as well.

Perhaps what's at issue is very simple: contemporary culture is numerically and scientifically illiterate. This endemic innumeracy is a widespread condition that allows some to use statistical data in ways that confound their audiences or that their audiences cannot challenge, let alone learn to "read" properly, according to Gerd Gigerenzer (2002). Here we are dealing with a broader educational deficiency rather than with the deliberate fabrication of data or public relations manipulations. But this still doesn't license some technoscientists who are experts at dealing numerically with data to exploit their expertise and confuse the

DOI: 10.1057/9781137519429.0006

public. Nor does it license their lawyers and lobbyists to discredit data of others as if they are junk. As Gigerenzer explains, whether we are dealing with breast cancer cases, AIDS, or DNA fingerprinting, opportunities to exploit the data are abundant, but there are simple lessons that can be mastered to counter such exploitation and ensure accurate representation of the data.

The concept of junk science, as we have seen earlier, has been used in order to discredit critics and reconcile "pro-corporate bias" (in courts of law as well as in the media) with "pretensions of scientific superiority" while dispensing with either controversial scientific issues or ethical conflicts of interest (Rampton and Stauber, 258–265). Mathematical illiteracy is not only a dangerous cultural condition (when dealing with technoscientific issues), but also a condition that gives rise to ethical hazards. Under such legal and commercial pressures, is there any hope to resort to the ethos of science of yesteryear? Can its ethical aspirations be resurrected? I would think that the practices of open-source code writing – as opposed to the patented intellectual property that some code writers secure – is a way to mitigate some of these concerns and demonstrate that corporate culture, with its neoliberal ideology, doesn't reign supreme.

Just as some in the scientific community have been complicit in the nationalistic tendencies of democratic regimes, accepting funding for warfare projects directed by the state, so have some in the same community been part of the commercialization of scientific research. It has been the case that the sheer size of some of these projects requires appeals to the state and/or to commercial interests for funding; even richly endowed universities cannot underwrite large projects. As we have seen earlier, the state can be excused for its nationalist needs for security and international prowess; there are always foreign threats (real or not), and a defense apparatus that needs to be bolstered with the aid of technoscientific ingenuity. However, the same cannot be said of the commercial interests we have briefly examined. Supplementing state funding with commercial grants isn't simply an economic strategy; instead, it bespeaks of shifting from science as a public good to science as intellectual property, within and outside the university system. And this shift is observed in the changing expectations of scientists and the compromises to which they lead. Under what institutional conditions can we promote countermovements that believe in public goods and in a scientific commons that should benefit society as a whole? Partial answers to these questions will be provided in the concluding chapter.

DOI: 10.1057/9781137519429.0006

4

Big Pharma: Pharmaceutical Dominance of Science

Abstract: *This chapter focuses on the deliberate misrepresentation of scientific facts by Big Pharma and the devastating consequences that affect an unsuspecting public. This kind of fraudulent activity forces scientists to make compromises to the ideals of science of a different and more pernicious kind than those made for national security or the marketplace. The central role the health-care industry plays in contemporary culture and our well-being makes this area of research and development of particular urgency for critical evaluation.*

Sassower, Raphael, *Compromising the Ideals of Science*, Basingstoke: Palgrave Macmillan, 2015.
DOI: 10.1057/9781137519429.0007.

Unlike scientists who set their sights on learning the mysteries of nature, there are those whose intervention with and modification of natural phenomena are predicated on definite financial gains. The expectation of this sub-group of scientists is narrowly focused on potential financial advantage and the control that their findings can attain. This form of control depends on either defining a new medical condition that requires a new treatment or prescription (various forms of depression and anxiety are regularly cited) or reclassifying a known drug that was effective for one medical condition as able to fight a newly diagnosed disease (as the examples later will illustrate). Perhaps the inspiration for this kind of scientific behavior can be related back to the horrors of eugenics (from the US to Nazi Germany). Originally understood as a means to bring about a more "perfect" human race (where cognitive and physical impairments are eliminated), this rationale for state intervention was eventually exploited by Nazi policies to exterminate millions of people so that only the Aryan race will dominate the world. When applied to the food industries, a whole range of Genetically Modified Organisms may help fight global hunger (as Monsanto has been claiming for decades) or may create genetically modified species the modifications of which may bring about changes whose long-term effects are yet to be determined. When applied to the pharmaceutical industry, new drugs and therapies are offered where none seems to be needed. Treatments are heralded for a newly codified set of behaviors as "sickness" or "illness," inventing cures for what shouldn't be problems to begin with. Most pronounced in drugs for psychological conditions, such as stress and anxiety, this multi-billion market has subverted the very activity of some scientists.

Routine scandals

One way to examine the prevalent abuse of scientific practice in the pharmaceutical industry is to follow reports in popular media about legal settlements by this or that corporation. Just to appreciate the magnitude of the economic impact of multinational pharmaceutical giants, let's review their 2013 global revenues (rounded figures are all in billions): Johnson & Johnson: $71; Novartis $58; Roche $47; Pfizer $52; Sanofi $45; GlaxoSmithKline $44; Merck $44; Bayer $26; AstraZeneca $26; and Eli Lilly $23 (Palmer and Helfand 2014). The top ten pharmaceutical

DOI: 10.1057/9781137519429.0007

companies have sales of about $436 billion globally; one can speculate that this segment of the global economy has about double the annual sales reported earlier, if one were to include all health-care related products and services. According to Marcia Angell, "The most startling fact about 2002 is that the combined profits for the ten drug companies in the Fortune 500 ($35.9 billion) were more than the profits for all other 490 businesses put together ($33.7 billion)" (2004, 11). Though outdated, this calculation is most likely still true for more recent years. When reviewing this industry, two related issues become apparent: on the one hand, by its very nature, this industry is focused on profits from sales, driven as all other corporate entities are by shareholders' ongoing thirst for growth and profitability in the name of curing the world's ills or bringing improved health conditions to every corner of the globe; on the other hand, this is an industry that explores the depths of human biology and health, inspiring technoscientific research in a wide spectrum of basic and applied fields worthy of funding.

Can these two interrelated facets of the industry live side by side without necessarily bringing about the breakdown of scientific ideals for the sake of maximizing profits? Some troubling reports come through the media suggesting that, as *The Economist* reports: "scientists like to think of science as self-correcting. To an alarming extent, it is not" (2013). Citing cases where in only 6 out of 53 studies or only one quarter of 67 experiments where test results were replicated by entities other than those originally reporting them, the article sheds light on one of the cornerstones of scientific self-policing, namely, the ability to replicate *all* test results. Likewise, the article cites the increase in retractions of experimental results, but admits that no more than 0.2% of annually published papers are being retracted, which is a small and reasonable fraction of all published articles. And finally, like many other governing boards of professional journals, there is a general awareness of the flawed peer-review process. Two responses are in order: first, this very article is itself part of the self-policing and self-correcting process that increases awareness of problems in the system and by reporting them makes an effort to address these problems; and second, that a certain level of "tacit knowledge," acknowledged by Michael Polanyi (1966), is bound to make replication difficult (because not all the variables are fully accounted for). Certain minuscule operations are built into the apparatus of experimental practices that are impossible to outline in the literature, and thereby may cause difficulties in replication.

DOI: 10.1057/9781137519429.0007

These concerns point out that some fundamental *trust* is being undermined among working technoscientists whose allegiance is now split between a corporate funding apparatus and the scientific community in general. Intellectual property rights and the patenting values they represent can overshadow the collegial atmosphere of bygone times, as enumerated in the various accounts of the ethos of science. When marketing goals are front and center, sharing basic data becomes untenable. Data are routinely presented so as to appease management demands for quick and successful results, suppressing inconvenient truths along the way or selecting small samples for tests. This is true not only when researchers are on the payroll of this or that pharmaceutical company, but also in the case of professional organizations and journals whose sustainability wholly depends on the largesse of these corporate giants (when they advertise or sponsor them). What conflicts of interest are they willing to overlook? What shortcuts are they willing to take in order to ensure their survival (DeAngelis 2006)?

Whether or not there is a will and a way to curtail outright abuses on the institutional level remains to be seen. In the meantime, scandal headlines fill print media and the airwaves. Katie Thomas reports that "Johnson & Johnson has agreed to pay more than $2.2 billion in criminal and civil fines to settle accusations that it improperly promoted the antipsychotic drug Riserdal to older adults, children and people with developmental disabilities, the Justice Department said Monday." If we think this is an isolated case, we find out that "Johnson & Johnson was not the only company marketing drugs to older dementia patients and the long-term care facilities where they were treated. Within the last five years, federal officials have reached similar agreements regarding Zyprexa, made by Eli Lilly; Seroquel, made by AstraZeneca; and Depakote, by Abbott, which is now AbbVie" (2013a).

Not long after this announcement, another headline appeared in *The New York Times* suggesting that "Doctors Say Heart Drug Raised Risk of Attack" (Pollack 2013). In this case, Anthera Pharmaceuticals refused to turn over data that proved not only the inefficacy of its drug to treat patients with acute coronary syndrome, but also that this drug could cause more harm than if not used at all. At issue here is sharing the data of the failed tests with academic researchers and clinicians at the Cleveland Clinic. Whether the trial proves to be successful or not, whether the study is stopped when danger is reported or continues, are issues that cannot remain internal to this or any other pharmaceutical

DOI: 10.1057/9781137519429.0007

company. Unfortunately, unless and until civil or criminal suits are brought against violators, one is left to wonder how to prevent such abuses in a more systematic fashion.

On the same day that Anthera Pharmaceuticals was mentioned, Johnson & Johnson was also reported in the media, this time settling a lawsuit about hip implants. The cost in this case was close to $2.5 billion, and would cover about 8,000 patients. What is outrageous about this case is that in a 2011 internal document, the DePuy Orthopaedics division of Johnson & Johnson estimated that the device would fail in 40 percent of the patients within five years (Meier 2013). Knowing that, one might ask, why continue? Why not improve the device and ensure long-term viability? It wouldn't take much to be proactive and offer replacements rather than await trial and pay fines, and in the meantime have thousands of patients suffer.

Within a month, we find another alarming headline: "Glaxo Says It Will Stop Paying Doctors to Promote Drugs." In the article, Katie Thomas clarifies that there are three issues Glaxo is willing to stop: not pay doctors to promote drugs, not tie their own representatives' compensation to the number of prescriptions doctors write, and not provide direct financial support for doctors to attend medical conferences, "common industry practices that critics have long assailed as troubling conflicts of interest." All of this comes after the fact that in 2012 Glaxo already paid a record $3 billion in fines because it had sold drugs for unapproved uses (2013b). One must wonder if paying annual fines has become a common industry practice. Are there no deterrent penalties that would affect changes in the pharmaceutical industry's procedures and policies? Or, are members of this capitalist elite immune from criminal punishment? Perhaps Glaxo's suggested "reforms" are telling: it will "continue to pay doctors consulting fees for market research" as well as continue to provide "unsolicited, independent educational grants." George Orwell would be proud that his fictionalized "doublespeak" of *1984* is being practiced in 2013.

That the pharmaceutical industry collectively behaves like any other capitalist industry is of no surprise. What is also of no surprise is the fact that profit motives overtake any ethical consideration in this capitalist framework (despite Smith's concern for moral sentiments, as mentioned earlier; see also Sassower 2009, Ch. 1). Finally, it also shouldn't surprise us that the legal system remains somewhat inept at dealing with – reigning in, critically evaluating, and properly punishing – medical science, whether in the case of breast implants (Angell 1997/1996), or any other

DOI: 10.1057/9781137519429.0007

that requires technoscientific expertise. Perhaps this industry is as much in the news because of its size and enormous profits, as it is about the fact that its fraudulent activities can kill innocent patients. This is also true about the asbestos and agricultural industries, as was seen earlier. So, what makes Big Pharma different? Just like the invasion of GMOs into the lifestream of developed and underdeveloped countries, so has the dominance of pharmaceutical companies been felt in a wide spectrum of drugs, devices, and treatments as health care becomes dominant in national expenditures (accounting for the single largest area of national expenditure). GMOs have changed the basic building blocks of the foods we grow (and consume) and the livestock we feed (and eat), and in this sense changed nature. Though more conceptually than a complete material intervention in the human building blocks, pharmaceutical companies (and their technoscientists) are changing, bit by bit, the nature of our physical and mental health. We turn now to the extreme cases whereby some corporate scientists have been instrumental in manufacturing illnesses and thereby changing what has been traditionally expected of scientists.

Pernicious tampering with nature

Among the first critics to rail against the presumption of "mental illness" was Thomas Szasz, himself a certified psychiatrist and professor of psychiatry. Following Sigmund Freud (1856–1939) and labeling their approach "scientific," mid-20th-century American psychiatrists and psychoanalysts were glad to adopt a cult-like religious fervor toward their approach. Szasz's critique became all too familiar 50 years later: "whereas in modern medicine new diseases were *discovered*, in modern psychiatry they were *invented*. Paresis was *proved* to be a disease; hysteria was *declared* to be one" (1974/1960, 12; italics in the original). The invention and declaration of mental illnesses is left to experts whose judgment is accepted by the public without complaint, because these experts are considered arbiters of the latest studies in neurology and other branches of medicine (Ibid. 47, 103). The invention of diseases isn't the exclusive purview of psychiatry, as Angell reminds us. Instead, this is a widespread *strategy* of pharmaceutical companies: "Once upon a time, drug companies promoted drugs to treat disease. Now it is often the opposite. They promote diseases to fit their drugs" (2004, 86). When wholesale

DOI: 10.1057/9781137519429.0007

invention is too outlandish, other *tactics* work just as well. For example, an arbitrary change in the definition of what counts as "high blood pressure (hypertension)" from above 140 over 90 all the way to 120 over 80 as "prehypertension" requiring medication; similar change in cholesterol measurement, from over 280 milligrams per deciliter to as low as 200 (Ibid. 85).

Peter Breggin continues this line of critique by pointing out a more precise tipping point for the transformation of the pharmaceutical industry to the time when President Bush signed a congressional resolution "declaring 1990 the first year of the Decade of the Brain. This has become psychiatry's main *promotional theme*, aimed at *selling biological and hospital psychiatry to the public* and at garnering more money for research into the brain as the seat of personal, social, educational, and political problems" (1991, 11; italics added). Mass-media went along with this promotional campaign, agreeing that "schizophrenia is a biological disease rather than a crisis of thinking, feeling, and meaning" (Ibid. 24). The fact that the literature supporting a genetic cause for schizophrenia, for example, has been reduced substantially over time seems to be glossed over if not ignored by the same media outlets that originally reported on the issue. Old studies that examined a "genetic basis" have been "discredited by the hundreds, while new ones are rare indeed" (Ibid. 99). So why is there still interest in making such claims for the scientific basis of psychiatry's classifications?

According to Breggin, two interrelated reasons motivate this yearning (rather than reality): first, it's about identity and ideology: "Their entire professional identity depends on this ideology, and in the case of researchers, their funding can be totally dependent on it" (Ibid. 94); and second, it's about professional politics, "the wish of psychiatry to maintain a medical image, to uphold its dictatorial authority, to garner federal funds, and to convince the patients to seek psychiatric help. *Psychiatry has tried to make depression into a political issue in America,* much like poverty, unemployment, or AIDS" (Ibid. 148; italics added). Though there are a few doctors here and there, like the Swedish Lars Martensson, who advocate "the right to drug-free care" (Ibid. 89), the vast majority of psychiatrists and psychologists comfortably follow the mechanistic model of medicine, when cells and genes have become the units of analysis. The mechanistic model, by contrast to other models, is more linear and deterministic insofar as it suggests that physiological effects can be reduced to specific and identifiable causes, and therefore

DOI: 10.1057/9781137519429.0007

lend themselves to treatment – intervention in the chain that leads from causes to effects – by drugs. It should also be noted here that psychiatrists who receive their degrees and certifications from medical schools are licensed to prescribe drugs whereas other clinical therapists cannot.

Breggin calls the move to finding a biological basis for mental conditions "biomythology," namely, a way of treating depressed patients as "biochemically defective mechanisms." The explanatory power of this move is indisputable for marketing purposes, both as a way to explain that people always feel anxious and that this anxiety can be treated if not eliminated with drugs. Following the classic economic axiom that "advertising actually *creates* consumer needs" (Jean-Baptiste Say (1767–1832) argued that supply creates its own demand in the marketplace), Breggin argues that by targeting those who are anxious, psychiatry can increase the demand for treatment and drugs. The sad conclusion is that the power of the "psycho-pharmaceutical complex" undermines criticism, so that the public remains unaware of alternative approaches to treatment (Ibid. 240, 365). Incidentally, a similar sentiment is voiced by Krimsky who complains that "significantly more research is done on the uses of chemical pesticides as opposed to biological pest control. Likewise, vastly more resources are put into the cellular and genetic basis of cancer than into environmental factors" (2003, 78). It's as much about what is transmitted to the public as it is about what is omitted that illustrates the subversion of science from its ethos of open-ended and critical inquiry.

The fact that sociological and political influences dictate a certain approach to health care in general and to medicine in particular can also be seen in the case of AIDS. Rebecca Culshaw's provocative title, *Science Sold Out: Does HIV Really Cause AIDS?*, says quite a bit. She starts by claiming that "AIDS has become so mired in emotion, hysteria, and politics that it is no longer primarily a health issue" (2007, 4). What kind of selling out does she have in mind? "To put it plainly, HIV science has sold out to the epidemic of low standards that is infecting all of academic scientific research." This means that "academics (young ones, in particular) are pressured to choose projects that can be completed quickly and easily, so they can increase their publication list as fast as possible. As a result, quality suffers" (Ibid. 14). She also reminds us that "science is not a democracy" (Ibid. 18) and therefore just because most researchers claim one thing and refuse to listen to their critics, their claim isn't necessarily valid (majority rule shouldn't dominate controversial questions in scientific research).

DOI: 10.1057/9781137519429.0007

The fascinating tipping point in this case isn't a presidential declaration or an enactment of law: "It was sometime in 1985 that HIV mysteriously went from 'the virus associated with AIDS' to 'the virus that causes AIDS,' squelching debate in the scientific arena. What changed? What happened to make scientists come to such certainty? If you look at the actual papers, you'll see quite clearly that the answer is: nothing" (Ibid. 19). And when drug studies are conducted, instead of comparing new drug therapy to previous drug therapies, placebo effect is used and that always shows positive results for new drugs by comparison (Ibid. 28–29). Just like Breggin and Krimsky, Culshaw provides a good explanation of retrovirus behavior as either exogenous or endogenous, and the refusal to explore environmental factors in favor of internal cellular ones, where drugs are more readily available (Ibid. 53). In summary, similar to psychiatric labels and classification, "accumulated data from years of testing indicate that the levels of HIV in the population are unchanging geographically – always higher in the East and the South than in the West and Midwest, unchanging in number, and far too consistent over racial groups and gender to be consistent with the irregularities of AIDS in the population. All the epidemiological evidence to date strongly indicates that whatever testing HIV-positive signifies, it clearly is not a reliable indicator of the risk of ever developing AIDS" (Ibid. 57).

In case the point here remains unclear, it should be emphasized that conclusions about infection may be reached prematurely. AIDS describes symptoms (CD4 or WBC levels) and HIV is the actual virus. One can be infected with AIDS without HIV (a better phrase is late onset of autoimmune deficiency) just as someone can have sepsis without liver failure or hepatitis. Discerning what can be known from medical tests from what is assumed to be the case requires careful analysis, so that fact can be separated from imagined fictions about a patient's condition. Medical professionals aren't confused about the details of their craft, though media outlets or interested parties, as described here, are more than happy to prey on such potential misunderstandings.

The language of the scientific method is alive and well, used rhetorically for marketing purposes, and at times it shifts public perception in the direction favored by pharmaceutical companies that fund research and sell drugs. Though confessional in tone, Katherine Sharpe's review of the impact of Zoloft follows this line of critique. She reinforces Breggin's indictment of psychiatry's shift to biomythology, accounting for the cultural acceptance and pharmaceutical shift to medicalize

DOI: 10.1057/9781137519429.0007

(rationalize in Weber's terms) mental health since the 1980s. Though originally developed as a blood pressure medication in 1987, Prozac (which contains SSRI – selective serotonin reuptake inhibitors) and other drugs like it have surpassed by 2005 "blood pressure medications to become the most-used class of drugs in America, with 10 percent of adults taking them in any given month. By 2008, that figure had jumped up to 11 percent" (2012, xiv–xv). This phenomenal growth is tied to a change in the cultural perception of depression and the ways by which it can be treated.

Classical views of depression include two sources: endogenous depression or vital depression which was internal and rare, and depressive neuroses which resulted from external circumstances. Over time, there was a radical change in the conception of depression, making the rare cases the new biomedical standard, taking over from the traditional Freudian model of psychoanalysis which came into disrepute by the 1970s (Ibid. 32, 37). The Diagnostic and Statistical Manual of Mental Disorders (4th edition) defines depression as "the presence 'most of the day, nearly every day,' for two weeks or more, of at least five from the list of nine symptoms that include 'depressed mood,' 'loss of interest or pleasure,' unintentional weight loss, sleep disturbances, psychomotor agitation or retardation, fatigue, feelings of worthlessness, diminished concentration, and thoughts of death or plans for suicide" (Ibid. 30). As the classification is so broadly inclusive, it allows for the diagnosis of a much larger portion of potential mentally ill patients. This expansion, of course, doesn't necessarily discount or dismiss all forms of depression as artificially constructed. There are many cases where simple chemical imbalance exists and therefore a chemically induced intervention can and in fact do improve one's outlook on life.

The broadening of the diagnostic net is coupled, according to Sharpe, with the rise of the biomedical model of mental illness, which suggests that mental disorders can be treated as if they were discrete physical ailments, accompanied by identifiable biological causes. But this view hasn't captured the American psyche on its own; instead, pharmaceutical companies spent millions of dollars on "initiatives to educate people about depression; these invariably drove home the message that depression is a chemical imbalance best treated with a drug that acts on chemicals." Along the way, psychoanalysis was discarded in favor of psychopharmacology. And with psychopharmacology, scientific terminology has been sought after. "The phrase 'chemical imbalance'

DOI: 10.1057/9781137519429.0007

gestures at the truth," we can all admit; but this convenient phrase also glosses over or conceals "all that we don't know, as well as the quotient or subjective reasoning that plays a part in any discussion of mental disorder" (Ibid. 43–47). Direct-to-consumer advertisement (over $2 billion annually in 2013), primarily through television and Internet channels, has the added effect that viewers begin to diagnose themselves, and once they do so, are likely to demand from their doctors drug treatment when such treatment may not be necessary at all.

This simplification on behalf of a "silver bullet" drug isn't all pervasive, of course, and exceptions can be found among mental health professionals and patients alike. As Sharpe admits, some clinicians speak of "a 'biopsychosocial' model of mental disorder, one that appreciates the interrelated contributions of genetic, psychological, and environmental forces" (Ibid. 48). But these voices are rare and weak. As critics have argued, "the 'medicalization' of what were once regarded as negative feelings or nuisance parts of life has harmed us" as people, as members of families, and as members of a community. Because "mental disorder is now over-diagnosed and psychiatric medications are overprescribed," we seem to be waging a "pharmaceutical warfare on ordinary sadness – a war that has given undue power to 'experts,' lined the pockets of pharmaceutical companies, and left the rest of us feeling enfeebled, more ill than we truly are" (Ibid. xix–xx). The damage of over-medicating with Prozac and Zoloft or any other antidepressant drug is one for generations to come. The war on sadness, as it has been labeled, may have as much to do with joblessness, low wages, and other cultural and environmental problems as with one's internal feelings and thought processes. And when public concern is voiced about a particular drug, a replacement one is right around the corner, one with less side effects or negative publicity.

This has been the case with the marketing of Attention Deficit Hyperactive Disorder, whether to children first and later to adults. As Alan Schwartz reports, "the number of children on medication for the disorder had soared to 3.5 million from 600,000 in 1990" (2013). Successfully publicizing the syndrome, pharmaceutical companies were able to reach doctors, educators, and parents and sell Adderall and Concerta to treat millions of children. ADHD has become "the second most frequent long-term diagnosis made in children, narrowly trailing asthma" with sales of stimulant medication reaching close to $9 billion in 2012. As problematic as "depression" has been to define and diagnose before pharmaceutical marketing efforts, so has been the case of testing for ADHD, leaving its symptoms to be vaguely interpreted by patients, parents, and doctors alike. Marketing campaigns, though, were effective even when misleading about the medical benefits of the drugs. Despite lax regulations since the 1990s, the FDA has systematically asked pharmaceutical companies to withdraw some ads, and even fined one company $57.5 million in February of 2012 (Ibid.). But Big Pharma is relentless, and the latest twist in its marketing strategy has been to target adults who have never before been diagnosed. In 2012 alone some 16 million prescription were written for people between the ages of 20 and 39. Will the next target market be senior citizens, those suffering from dementia? And in case they forget to take their medicine, new smart-phone applications are available for daily reminders, bells, whistles, and all.

The point to be made here is that Big Pharma not only invents diseases when none is apparent, it also attempts to grow its markets by targeting patients who shouldn't be patients at all, using clever and seductive advertising campaigns that promise to solve every physical or mental problem with the easy intake of daily pills. Shannon Brownlee cites the case of Merck's Vioxx which had no beneficial effects, sold billions of dollars of pills, and had some negative side-effects; the FDA as a regulator failed. (2007, 211) So, how is it that drugs like these find their way to the market? Brownlee reminds us that in 1992, "under pressure from the industry, Congress passed the Prescription Drug User Fee Act [which] permits companies to pay fees to the FDA in order to speed up drug approval – creating in the process the absurd situation in which the agency is now partially funded directly by the industry it is supposed to regulate." (Ibid 218) The fees paid by companies could be viewed as bribes or the lubricators of faster and less critical scrutiny of proposed drugs; FDA defenders argue, by contrast, that the agency is underfunded and therefore cannot handle the number of cases before it in a timely fashion, and therefore the fees are merely necessary for the smooth operation of the agency (without thereby skewing testing results or professional judgments). Despite the rationale for such intimate relations between regulators and those they regulate, there is an obvious conflict of interest here, so blatant in fact that scientists and regulators are equally condemnable in their merry-go-round of switching positions every few years, at times being researchers, at others the regulators of their previous research colleagues.

DOI: 10.1057/9781137519429.0007

Brownlee emphasizes the development of "a new lifestyle drug," a drug that is supposed to enhance whatever one wants out of life as opposed to treating an ailment. He mentions "Prozac for depression, Claritin for allergies, or a drug like Lipitor that treats a risk factor like high cholesterol," but he could have also mentioned Viagra (Pfizer), Cialis (Eli Lilly), or Levitra (Bayer), the erectile dysfunction drugs for older males whose libidos have passed their prime and whose blood flow may be compromised. The sales of these drugs have collectively reached $4.5 billion in annual sales by 2013. These drugs are cash cows because they can "be taken every day by huge populations, sometimes for years on end," thereby ensuring steady income streams for pharmaceuticals (Ibid. 218–219). This means that companies create entirely new markets for "me-too drugs" by promoting new sicknesses, like social anxiety disorder or stretching the definition of terms such as depression, stress, and anxiety. It should be clear, though, that not all the drugs listed here are of the same kind: some in fact relieve or treat physiologically traceable conditions, whereas others may have only limited psychosomatic effect. The efficacy of this strategy depends to a large extent not only on television advertisements, but also on convincing clinicians to market these diseases and drugs to their patients and each other (Ibid. 219–220). The mutual reliance on doctors' recommendations, paid for by drug manufacturers either directly (promoters, consultants) or indirectly (conference attendance), keeps the industry insulated from external scrutiny or the wrath of weakened regulators. Mirowski calls these corporate giants "the Madoffs of the modern commercialization regime" (2011, 205), leading him to conclude that concerns with outlying fraud should be understood institutionally and structurally, rather than the aberration of morally challenged individuals (Ibid. 229).

The outrage over the over-prescription of drugs and the lack of regulatory oversight are indicative not only of political and moral failures that are institutionalized (and thereby indirectly sanctioned), but also of the underlying scientific issues that become more significant than one would think. The basic question is whether or not drugs like the ones descried earlier are effective, and if yes, to what degree, and in relation to what specific population. The determination of these issues can be empirical and can be made public (just as is expected from the adherence to the scientific ethos). The public can then easily discern for itself under what conditions these drugs should be used and by whom, rather than being reassured that professional guardians are ensuring health-care safety (considered paternalism and morally problematic).

DOI: 10.1057/9781137519429.0007

Technoscience revisited

By now it's clear that the *scientific community* has become a *scientific enterprise* where monetary goals overwhelm the scientific ethos discussed earlier. But does this necessarily mean that moral codes of behavior, especially in the health-care arena, should be ignored? I suggest that it's important to recall some of the historical background in the 20th century to assure ourselves that not all is lost, and that legal recourse is not based on the goodwill of this or that judge and jury, but on a deep-seated cultural commitment to protect patients against abuses, however inexpedient (and at times costly in the short run) it may appear to individual doctors, medical administrators, or corporate and hospital administrators. The claim that individuals should be responsible for their own health or the conditions of their well-being is misguided; there are objective health (genetic) and environmental conditions over which the individual has no control. In these cases, expecting personal responsibility as a pre-condition or a motivation for individual life-style changes (diet and exercise, for example) remains untenable.

After World War II, Nazi doctors were put on trial by the US military under the authorization of President Truman. Beginning on December 9, 1946 and ending on August 20, 1947, the judges found that these doctors were involved in millions of sterilizations of German citizens, and in human experiments on thousands of inmates in concentration camps. Few doctors were found guilty, and punishment was usually disproportionately minimal in light of the gravity of the crimes. Leo Alexander submitted to the Counsel for War Crimes six points defining legitimate medical research to which four more were added, constituting the "Nuremberg Code" that eventually informed the legal framework both in the US and most of the world. The Nuremberg Code includes such principles as informed consent and absence of coercion; properly formulated scientific experimentation; and beneficence toward experiment participants.

Perhaps quoting the first principle (out of ten) would suffice here to highlight the point being made in regard to the cultural (and legal) framework within which health care should be practiced:

> The voluntary consent of the human subject is absolutely essential. This means that the person involved should have legal capacity to give consent; should be so situated as to be able to exercise free power of choice, without the intervention of any element of force, fraud, deceit, duress, over-reaching,

DOI: 10.1057/9781137519429.0007

or other ulterior form of constraint or coercion; and should have sufficient knowledge and comprehension of the elements of the subject matter involved as to enable him/her to make an understanding and enlightened decision. This latter element requires that before the acceptance of an affirmative decision by the experimental subject there should be made known to him the nature, duration, and purpose of the experiment; the method and means by which it is to be conducted; all inconveniences and hazards reasonable to be expected; and the effects upon his health or person which may possibly come from his participation in the experiment. The duty and responsibility for ascertaining the quality of the consent rests upon each individual who initiates, directs or engages in the experiment. It is a personal duty and responsibility which may not be delegated to another with impunity. (United States National Institutes of Health)

The rest of the principles warn against undue risks, the potential for more benefits than costs, the elimination of unnecessary physical and mental suffering and injury, the significance of the experiment, adequate preparation and safeguards for patients, and the appropriate qualifications of scientists.

When revelations of the Tuskegee Syphilis Experiment came to light, the horrors of Nazi abuses were invoked and revisited. Apparently, the US Public Health Service conducted between 1932 and 1972 a study on the natural progression of untreated syphilis in rural African- American men who thought they were receiving free health care. The experiment included 600 impoverished sharecroppers from Macon County, Alabama. Of those men, 339 had contracted syphilis before the study began, and 201 did not have the disease. The men were given free medical care, meals, and free burial insurance for participating in the study. They were never told they had syphilis nor were they ever treated for it. According to the Centers for Disease Control, the men were told they were being treated for "bad blood," a local term for various illnesses that include syphilis, anemia, and fatigue. Why were these men not treated when treatment was available? Why weren't they informed about their condition or the terms of the study? Why was the Nuremberg Code not enforced after its adoption in 1947? One wonders what racist or socio-economic biases informed such an unethical human experiment (and probably illegal because of civil rights violation) as late as 1972.

A similar revelation, years after the fact, came to light in the case of Henrietta Lacks whose cancerous cells were reproduced for generations without her knowledge and consent or the knowledge and consent of her descendants. Rebecca Skloot published the results of her investigation

DOI: 10.1057/9781137519429.0007

about what happened in 1951 to a poor, African-American, young woman, and the successive revelations that were known in the circles of genetic researchers but unknown to the family of the patient or the public at large. Though the term "informed consent" didn't find its way into the court system till 1957, there were already then questions about how much to "inform" the patient and what "consent" would mean for a patient under duress (2011/2010, 132). The Nuremberg Code may have been formulated already in 1947, but its adoption in medical schools and health-care facilities would take another decade or two with Institutional Review Boards for human experiments (1974).

One would hope that by the 21st century these kinds of ethical questions and failures would be clearly demarcated in all technoscientific institutions and among all those who are trained to undertake research (regardless of neoliberal ideological or corporate pressures). But such codes or even legal precedents remain background noise that can be, and at times is, sadly ignored in the face of the real sound of money. When Szasz, for example, reminded us of the pernicious impact psychiatry might have on the treatment of patients and our culture in general, he was worried about the wholesale abdication of common sense in the face of medical expertise and the demands it makes on our lives and pocketbooks. His warnings may have been ignored over the past 50 years, but they should be heeded when we observe how over-treated we have become by now. We are not necessarily becoming healthier because of over-diagnoses and over-prescription of drugs and procedures, but instead carrying the burdens of expensive, and on many occasions, unnecessary interventions that are dictated by the twin maladies of fear of malpractice suits and the incentives associated with expensive testing.

The ethical compromises of the health-care industry – from Big Pharma to hospitals and doctors – seem clear cut even when exceptions are found. One such exception worth noting in passing is the case of Jonas Salk (1914–1995) who discovered and developed the first successful inactivated polio vaccine, and who refused to have it patented. He insisted that the vaccine should be available to the public without him or any pharmaceutical company profiting from preventing a debilitating affliction. So, what remains for us to undertake here is a serious reconsideration of the conditions under which changes can be made so that the exceptions to the rule become the rule. Ethical codes by themselves will do nothing to change the culture of greed and corruption, nor will court cases that levy fines on those caught abusing the system (because

DOI: 10.1057/9781137519429.0007

for some it will simply be an additional cost of doing business). Instead, we must rethink the political–economic framework within which incentives to misrepresent the facts are substantially reduced. The false fascination with the 17ᵗʰ-century gentlemen of science and the uncritical adoration the popular images of genius scientists are powerful and make sense in this context, because they appeal to an honor system among technoscientists that would move them professionally to do the right thing, to adhere to the scientific ethos and all that it stands for. Perhaps in this spirit we have to focus on contemporary attempts to provide a community-wide adoption of the open-source system, as an example. Such experimentations may guide us as to how the basic tools of science can be offered to all potential developers regardless of their initial funding (eschewing intellectual property rights). Likewise, other contributions to our shared Commons (from code to space and beyond) would remind us that morality does matter in the digital world. I would venture to guess that Adam Smith's Impartial Spectator – whether understood as our collective and personal conscience or as a God-like presence – could now be invoked as a means to defy what has become super- or hyper-capitalism. Perhaps there is room for radical transformation not despite, but because of how poorly technoscience has fared under the pressures to increase profits and ensure economic growth. The price we have collectively paid, in developed and underdeveloped countries alike, has been too high. As I'll attempt to illustrate in the next chapter, we can aspire to do better, and even manage to transform our outlook along the way.

DOI: 10.1057/9781137519429.0007

5
Situating Technoscience

Abstract: *In this concluding chapter, various suggestions and examples are brought to light where the ideals of science can be found or resurrected. Given the conditions of the postmodern digital age, there is a growing interest in thinking about scientific knowledge in broader terms than those of the state or the marketplace, realizing that we can organize our technoscientific research and development virtually and globally, and because of that, our responsibilities transcend our own local or national interests.*

Sassower, Raphael, *Compromising the Ideals of Science*, Basingstoke: Palgrave Macmillan, 2015.
DOI: 10.1057/9781137519429.0008.

DOI: 10.1057/9781137519429.0008

We must recall that the point of this book has been less the documentation of technoscientific fraud or the various ways in which technoscientists here and there compromise their adherence to the ethos of science, and much more about the fact that these cases are assessed against the backdrop of the bygone ideals of science. Perhaps what is more pressing at this juncture is whether the public's perception of an idealized technoscience is warranted or not in the postmodern digital age of the 21st century; likewise, we ought to be asking under what conditions some of these ideals of science should be flourishing again, even if they were never fully implemented or deployed by the scientific community. With this in mind, our judgments should be more nuanced, even forgiving to some extent, because setting up a lofty set of ideals no one is expected to fully follow is problematic on two counts. First, shouldn't the alleged absolute ideals themselves be regularly modified (because of changing circumstances)? And second, how can we hold technoscientists as members of communities and institutions personally accountable for ideals their communities and institutions themselves have revised or never fully adopted as their own (implicitly or explicitly)?

When considering these two interrelated questions, it should become clear that the soul-searching philosophers expect of their fellow academics and professionals may seem a luxury many practitioners cannot afford, even though the culture at large may want them to take the time to consider them. In this context, one can ask, does the public provide the means by which such soul-searching can take place? Is the public ready to pay – in time and money – for these reflective moments in seminars and workshops, for example, in light of productivity and time constraints? Or, as has been the case more often, does the public demand such an exploration only after disasters happen, from bridges collapsing to the explosion of the Space Shuttle Challenger in 1986? In an age characterized by some as postmodern – denoting a plurality of viewpoints none of which is inherently preferable to another, judging outcomes within prescribed contexts rather than in absolute terms, and remaining tolerant to the contingencies of life and the world without a prefigured goal or telos – it makes sense to remain open-minded and critically scrutinize technoscience in multiple ways. This manner of scrutiny suffers from lacking a foundation against which to set standards; but it enjoys a freedom of thought and imagination that may propel and excite future students of technoscience to excel and take risks they might not otherwise.

DOI: 10.1057/9781137519429.0008

Democratizing technoscience

With this in mind, I hope the rest of this chapter can provide some ways in which we can think about our changing expectations of technoscientists, as well as the conditions under which their research can be protected and allowed to flourish for its own sake. Underlying any practical suggestion is the ethos of the scientific enterprise as a moral benchmark. It's unrealistic that technoscience will ever be completely independent of its economic, social, political, and legal surroundings. But even when working within specific constraints, technoscientists can guard against and divert the changing expectations of their professional life, so as to maintain both personal and institutional integrity. Perhaps what might be the most important change in attitude would be a realization that no different from public intellectuals (Sassower 2014), technoscientists should think of themselves as *civil servants* whose responsibility is to the scientific commons and the overall well-being of society as a whole. Those amongst them that are more interested in fame and fortune should present themselves as entrepreneurs or businesspeople rather than technoscientists, just as academics who are consultants for the private sector should consider themselves entrepreneurs and businesspeople. I don't mean to condemn or condone one position over the other, just to clarify that one's intentions and goals should be spelled out so as not to confuse the public.

Instead of reverting back, in this concluding chapter, to the Weberian ideal, Mertonian ethos, or Nuremberg Code of scientific behavior, or any of the national guidelines proposed by government agencies, perhaps we should fully accept the era of *scientific enterprises* while refusing to concede to some of its unintended consequences. The shift to the notion of a scientific enterprise acknowledges the financial and other social contexts within which scientific research is undertaken; it also appreciates the community of scientists with its internal dynamics and the power relations that ensue within it; it fully appreciates the conditions under which it can or cannot minimize its dependence on specific commercial entities in order to further and sustain the autonomous pursuit of knowledge; and finally, it forces technoscientists to recognize and publicly declare what role they wish to play: *public servants* whose sole interest is the improvement of the human condition or *entrepreneurs* who wish to pursue fame and fortune. There are various benign consequences to all of these conditions, unlike the pernicious ones discussed

DOI: 10.1057/9781137519429.0008

earlier, such as the realization that one is not an island to oneself, namely, that membership in a community of researchers is a meaningful and responsible position, one that requires some attention to the rules of the game – the ethos of science – and a strong level of mutual respect and support. In what follows, I wish to highlight some of the ways in which technoscientific responsibility ought to be understood, especially when accounting for the widespread authority of technoscience.

One classic example of one's membership in a group or state (and its attendant authority) is the Milgram experiment conducted in 1961 at Yale University. The experimenters were trying to answer questions about one's obedience to figures of authority. The astonishing results of this experiment illustrated that students were willing to inflict pain on others to a much greater degree than was anticipated. Two standard interpretations of this unexpected behavior came out of these experiments: one was a "theory of conformism" that limits the authority and power of decision-making to experts or those higher in a hierarchy, so that those participating were willing to follow the orders of the white-robed experimenters regardless of how morally questionable they were; the other has been dubbed the "agentic state theory," which explains how people see themselves as obedient agents of the state, or figures of authority that claim to represent the state (Milgram 1974). Whether or not the subjects were right in following orders (as the experiment was a hoax – no electrical shock was ever administered) as Robert Shiller (2005, 158) argues, it seems that what is more relevant here is that they all were influenced, as Clifford Stott explains, by the idealism that scientific enquiry had on the volunteers: "The influence is ideological. It's about what they believe science to be, that science is a positive product: it produces beneficial findings and knowledge to society that are helpful for society. So there's that sense of science providing some kind of system for good" (2009).

Technoscience, in short, has the kind of privileged conceptual and practical status among contemporary members of our society that in its name and in the name of its ideals, we are willing to do things we would otherwise refuse to do. Its authority, then, is even more legitimate than that of the state. This legitimacy, to be sure, is couched in idealized terms already explained in Chapter 1, terms that themselves are idealistic and seem completely non-authoritative; they don't denote power relations, but instead the universal sharing of knowledge for the benefit of society. And this powerful institution is powerful not because of its police, military, or legal apparatus (as is the case with the state), but because of its

DOI: 10.1057/9781137519429.0008

ideals, the venerated ethos it claims to uphold. Obedience to this institution, then, denotes a reasonable and even commendable adherence to a set of transcendent principles worthy of human conduct. It is therefore not surprising that the participants in the Milgram experiment seemed more compliant than expected; their compliance was misunderstood in power-relation terms, rather than in idealistic terms we should hope to encourage.

The second kind of innocent unintended consequence of scientific inquiry that should be considered by those who wish to join this august community is illustrated by the tragic case of Fritz Haber. In 1909, he invented a synthesis of ammonia from nitrogen and hydrogen, in order to provide a better crop fertilizer, and later during World War I invented both pesticide and poisonous gases to be used to shorten the war. How tragic and ironic that Haber's inventions were eventually used by the Nazis under the Zyklon B label to gas millions of Jews, including his own family (Bowlby 2011). This case comes close, of course, to the kind of anti-fascist and patriotic arguments used during the development of the atomic bomb in World War II or later during the Cold War, but one whose tragic nature reminds us of the unintended consequences of technoscientific inquiry and application. This case is worth mentioning as an accidental, and not an institutional failure that might come about at any time under any set of circumstances. Given the open-ended nature of technoscience (from inception to application), should cases of innocent and misguided behavior be forgiven (however tragic their results)?

As a third case, we might also benefit from recalling Francis Bacon's warnings about the four idols that are bound to derail positive or altruistic scientific explorations. "The Idols of the Tribe" have to do with human nature and "the sense of man as the measure of things," a feature of scientific inquiry that therefore must be kept under check. "The Idols of the Cave" have to do with individual biases and prejudices related to one's personal disposition and education, and they, too, should be under scrutiny and avoided when possible. Whereas the first two are more personal, the second two are institutional. "The Idols of the Market Place" relate to the interaction among people and their communication, using words poorly to describe or assign truth values (what Foucault would call discursive regimes). "Idols of the Theater" have to do with how nature has been represented by tradition with all the errors associated with these representations over time (1985/1620, 48–49). These two institutionalized features of the scientific community must be explored

DOI: 10.1057/9781137519429.0008

and made explicit so as to eliminate errors they might bring about. If "rightly is truth called the daughter of time, not of authority" (Ibid. 81), it stands to reason, following Bacon, that fraudulent behavior among some participants in scientific enterprises will eventually be exposed, retracted, and excised from the scientific edifice. With time, their impact is reduced, if not outright eliminated.

Even when our society expects from technoscience its health and the well-being of the environment, it shouldn't blindly follow its pronouncements. The three cases listed earlier, from the specific ones of Milgram and Haber, to the more general one of Bacon's warnings, remind us to watch out for the powerful position technoscience occupies in our culture: its claim for authority accompanies its pronouncements, however carefully discursively couched or however reluctantly expressed publicly by the technoscientific community itself. The public is prone to accept more readily claims made in the name of technoscience than those of politicians and economists. This authority, however implicitly assigned or explicitly articulated, carries with it greater responsibility, both personally and institutionally. Neil Postman offers the following antidote to the alleged blind acceptance of the authority of technoscience: "a *technological resistance fighter* maintains an epistemological and psychic distance from any technology, so that it always appears somewhat strange, never inevitable, [and] never natural" (1992, 185; italics added). A level of vigilance is necessary to ward off Bacon's Idols and ensure a skeptical acceptance of technoscientific messages. As we have seen earlier, at times such caution can be an excuse for inaction (when it gets to pollution, for example), but at other times this antidote can save us from unnecessary medical drugs and procedures. Either way, this kind of skeptical and critical scrutiny clarifies the changing expectations of scientists.

The issue here is not simply providing safeguards against technoscientific malfeasance but a fuller appreciation of the democratic framework within which the scientific enterprise thrives. From philosophers and sociologists to policy wonks, it is obvious that the more the public is engaged in the affairs of the scientific enterprise the better. This way there'll be less of a dissonance between idealized and realistic expectations. Alan Irwin, for one, lauds The Association of Science Workers and the Royal Society who share a fundamental belief in the "centrality of scientific development to the future of society – and a belief (whether as part of a social democratic or more vaguely liberal ideology) that a better informed citizenry can play a crucial (but essentially reactive) role

DOI: 10.1057/9781137519429.0008

in this development" (1995, 14). When more options are available, when sources of knowledge outside the "enterprise zone" are legitimated, when alternatives are presented (Eastern healing techniques, for example, to treat pain), it's reasonable to hope that the fruits of technoscience will still be fully appreciated. In other words, more alternatives will not necessarily undermine the legitimacy of technoscience, only highlight its limits and the benefits that can be expected within these limits. The public then will ask "not *whether* science should be applied to environmental (and, of course, other) questions but rather *which form* of science is most appropriate and in *what relationship* to other forms of knowledge and understanding" (Ibid. 170; italics in the original).

Is this view of the democratic inclusion of the public in highly technical technoscientific debates too optimistic? According to Charles Thorpe and Jane Gregory it is: "contrary to the rhetoric of democratization that has accompanied public engagement efforts, these programmes [of participation] potentially operate as forms of control and co-optation, and promote the shaping of publics as markets" (2010, 273). We have encountered earlier the concerns with the economic rationalization of all decision-making processes in Western democracies in general terms, as they apply to "academic science," or the transformation of *science* into the *scientific enterprise*. And here are Thorpe and Gregory examining in detail some British forums that end up co-opting the public into the discourse of the scientific enterprise. When options are presented by industry, when choices are narrowly defined, what appears on the surface as democratic participation becomes rubber-stamped submission to prefigured processes.

Capitalism isn't fading away in Western democracies as quickly as some would like us to believe, nor is the technoscientific enterprise about to lose its luster or at least its shining gadgets. But this by itself should induce action rather than further concessions. We can follow Bacon and Babbage and warn our fellow citizens, offer ways by which to ensure the integrity of the scientific enterprise, and recalibrate our expectations of scientists. To begin with, as Krimsky reminds us, we should promote public-interest science, which he defines as "the rich tapestry of involvement that professors of universities have in providing expertise to government agencies and not-for-profit organizations in addition to pro bono help that many of them offer to underrepresented communities" (2003, 215). This would include public-interest science and an independent and critical voice within the scientific community.

DOI: 10.1057/9781137519429.0008

From *public engagement* in the broadest democratic sense we can move to the *critical engagement* of independently minded scientists and citizens. Perhaps greater national support for science for science's sake could be helpful, too.

The boundary conditions under which such critical analysis will become possible are outlined by Krimsky. His three principles directly address the roles different participants ought to play in the technoscientific arena. First, "The roles of those who produce knowledge in academia and those stakeholders who have a financial interest in that knowledge should be kept separate and distinct." This would ensure less immediate interference in research programs and the expected results of funders. Second, "The roles of those who have a fiduciary responsibility to care for patients while enlisting them as research subjects and those who have a financial stake in the specific pharmaceuticals, therapies, products, clinical trials, or facilities contributing to patient care should be kept separate and distinct." This is an obvious requirement already stated in the Nuremberg Code. Third, "The roles of those who assess therapies, drugs, toxic substances, or consumer products and those who have a financial stake in the success or failure of those products should be kept separate and distinct." This deals with regulators whose distinct responsibility is to protect the public, and should neither be funded by those they regulate nor be themselves researchers as well (Ibid. 227).

It's unrealistic to expect Big Pharma, for example, not to promote drugs and devices the way other companies market their wares. They all are trying to "sell" something and in so doing have a tendency to promise more than they can deliver. It's likewise unrealistic to expect that economic rationalization will disappear from contemporary cultural and political discourses, just as it is unrealistic to expect that the political apparatus will become independent from those funding its campaigns (as seen more egregiously in recent times in the US). But despite all of this, the beauty of science and the high esteem by which it is still held shouldn't be forgotten. Whether we can or cannot overcome inevitable conflict-of-interest situations remains to be seen. Discussing the issue openly may enlist some outrage and caution, and even contribute to changing the demeanor of those engaged in or critical of the scientific enterprise. Whether we can or cannot enforce modest changes in peer-reviewed publications will be seen soon enough. Judson, for example, cites an interesting proposal by Drummond Rennie to have contributors to scientific papers be listed the way movie credits are listed: each

DOI: 10.1057/9781137519429.0008

contributor is identified with the exact contribution s/he made to the project (2004, 310). This will go a long way toward acknowledging exactly what every person contributed to the research, instead of a list of 200 members of a laboratory where some worked very hard on the project whereas others did little or nothing. This minor proposal may help ensure greater transparency and accountability on the one hand, and a more accurate reward system.

Postmodern inspiration

On a larger canvas, we should paint an image of science as inspiring as any marketing campaign by for-profit enterprises. We should learn from public-relations experts how to sell a vision of technoscience rather than sell out the integrity that has always been its hallmark. In other words, *we should appropriate the tools of capitalism to subvert it*: with radiant and bombastic rhetoric we must capture the public's imagination of why curiosity and the quest for knowledge are worthy of our collective efforts and funding without any strings attached. We must highlight the international effort at CERN, for example, the European Organization for Nuclear Research, and remind the world that finding the so-called God Particle (The Higgs–Boson or Higgs particle is an elementary particle initially theorized in 1964 and discovered on July 4, 2012) was a collective effort by thousands of technoscientists from around the world. Collaborative work without immediate financial rewards is alive and well, so it seems in this case.

Likewise, we should highlight, as the documentary television program "60 Minutes" popularized on December 7, 2014, an effort by Patrick Soon-Shiong, a billionaire physician who is embarking on a quest for a new method for identifying and treating cancer. He has drawn the attention of the UK government and Oxford University on the one hand, and the attention of local pharmaceutical companies that already have had some success in genome sequencing, so as to study oncology differently from established approaches (according to him). Heralded by some as the "NASA of biotechnology" (Taylor 2014), the expectation here is that an innovative individual, admittedly a very wealthy one, has still the ability to inspire others to collaborate and perhaps find a personalized solution to cancer. To be sure, what this method of treatment wishes to highlight is how to match the individual's genetic makeup and oncological condition

DOI: 10.1057/9781137519429.0008

with a customized response (rather than the earlier method of "one size fits all"). Whether this effort is Quixotic or brilliant (and operational) remains to be seen. It's also unclear what the financial rewards will be for those who collaborate on this new vision of immunotherapies for cancer (Garde 2014). Yet this instance illustrates my earlier comments in Chapter 4 about the postcapitalist conditions with which we must still work: some will take this case as proof for yet another profit-motivated pharmaceutical push, whereas others may consider this case as a pull toward the value of an individual speculation and insight that may yield no results whatsoever. Should the billionaire be trusted? Does he get to test his hypothesis because of his wealth or because of his brilliance? However these questions are answered, we can agree that attempts like this one should be communally funded to ensure better results than just following one research path.

If the cases of collaboration for potential public benefit sound too utopian for some, there have been various attempts to square the circle, so to speak, of the ideals of science within and beyond the confines of capitalism. As I have outlined elsewhere (Sassower 2009), *postmodern postcapitalism*, which draws on the best both capitalism and socialism can offer, is agile enough to be contextualized in light of specific needs and resources and is in fact already practiced. In areas such as fashion and gastronomy, we are familiar with the *knockoff economy* that allows copying without the protection of patents and copyright (Raustiala and Sprigman 2012), just as we are familiar with *open-source* platforms that enhance the proliferation and success of code writing without copyrights (Berry 2008). We also see the growth of *crowd-sourcing* which, unlike venture capital and even angel investment, fosters a community of contributors whose only reward is seeing the accomplishment of a dream: the dreamer (initiator) isn't beholden in any way to funders' generosity, unless s/he wishes to give them a gift in return for success. Though limited in their scope, these postmodern and postcapitalist experiments are viable and sustainable. They speak to the skeptic who might view a philosopher's critique too utopian. They are also becoming more fashionable to promote, as we see in the case of Jeremy Rifkin (2014). Despite rigid ideological commitments of the right and the left, the so-called conservatives and liberals, a plurality of modes of production, distribution, and consumption are being tried everywhere. Yesterday's dream may be tomorrow's reality, after all.

They also speak to the need of re-conceptualizing the role of the state in regard to technoscience. First, a more democratic approach would

DOI: 10.1057/9781137519429.0008

necessarily redress some of the abuses prevalent with either nationalistic or corporatist encroachment into scientific practices, as Agassi (1981) and Robert Dahl (1985) have argued years ago. Second, the economic dimension of state investments in science is central, as Mazzucato has recently argued (2014); it isn't simply a conduit, arbiter, or mediator among competing technoscientific visions (with the attendant critique that it poorly picks winners that turn out to be losers), but instead the prime entrepreneurial force that accomplishes more than can be expected by a fairly conservative marketplace. To remain sustainable, the state could either impose a 1% fee on all Internet activities (originally funded by the state), set up state investment banks (already operational in some European countries) that would expect returns on their loans to continue future investments, or expect stakes in all enterprises relying on state funding (grants or loans) to ensure fair compensation. The risk–reward matrix can be easily modified to ensure a sustainable rate of investment in technoscience. Third, guiding these suggestions for reframing the future of technoscience is also the idea of *Pay It Forward* (Hyde 1999): inspiring altruistic acts of kindness whose repayment wouldn't directly affect the initiators, but complete strangers who in turn would do the same to others. Is this still too utopian? I do not know. But I do know that it is inspired by the postmodern mindset that is willing to entertain any novel idea and experiment with any proposal, as long as its own success and failure remain as part of its ongoing evaluation.

I could have devoted a chapter to Big Data as another area in which the expectations of scientific community have changed in recent years. Elsewhere (Sassower 2013b), I argued that despite the many changes brought about by digital technologies and the introduction of the Internet, their cultural framing hasn't been as radical as it may seem on the surface. That is to say that there are always critical voices that fear the introduction of new technologies (technophobes) and those who welcome them with open arms (technophiles). In an era of *postcapitalism* – a hybrid constellation of capitalism and socialism seems inevitable in the welfare state – pecuniary motives still remain, even when attenuated. These motives may shift from profit maximization to sustainability, yet they are still of concern and we need to address them. Likewise, human interaction, though characterized by speed, ease, low costs (Rifkin 2014), still remains bound by the cultural conditions of yesteryears – we are still worried about power relations and concepts, such as love and happiness, jealousy and depression. Utopian views of Big Data – from smartphones

DOI: 10.1057/9781137519429.0008

with numerous applications to profound new means by which to monitor the health of patients with embodied chips (Topol 2012) will always have to be balanced against their dystopian counterparts: connectivity breeds surveillance as much as some collaborative commons cannot eliminate human vices that lead, for example, to identity theft.

I don't mean to minimize the importance of digital technologies and their potential for offering wonderful intellectual and perhaps emotional fruits (of connectivity or immediacy, if not intimacy); instead, I wish to highlight the fact Big Data offers the same conditions for changing the expectations of scientists in the face of new realities and the compromises these changes will entail. Should technoscientists refuse any complicity with national surveillance of citizens? Should they insist on the openness of their discoveries and eschew intellectual property rights? These and many other questions are similar enough to those raised in the previous three chapters that they would be redundant in the case of Big Data. Surely some would argue that the digital age, just like the Industrial Revolution, has been a game changer and therefore deserves scrutiny. I don't fully agree with this sentiment, because such assessments may be premature, given that various unintended consequences of this age have yet to appear.

The inspirational speeches by the likes of President Kennedy about space exploration or James Killian, the chairman of the board of MIT, about the service science can offer, or those before the US Congress in the case of the Superconducting Super Collider bring to life the spirit of science. Perhaps we shouldn't pretend that scientists can overlook the temptations of fame and money or escape the clutches of corporate funding. Likewise, to think of technoscientists collectively as having lost their way and as fallen angels or whose activities are all tainted by nationalist and corporate interests is unfair. When we step back and look at the overall contribution of technoscientific research to humanity, we can forgive (if not forget) some minor indiscretions, whether preventable or not. At the same time, we should all strive for technoscience to have a higher moral stance, one that recalls the expressed ethos of science, and in this sense elevate the expectations of scientists to the level of angels, so to speak. The reason for this elevated expectation is that the consequences of technoscience in general are much broader and more fundamental than those, say, of philosophy or music. Grave suffering and pain can be inflicted with technoscience in a fashion more drastic than in many other human endeavors, and therefore the stakes are higher. If the ethos

DOI: 10.1057/9781137519429.0008

of science seems too lofty, perhaps we should remind ourselves the extent to which it's meant as an ideal, a heuristic that should guide our views and actions, and not as a set of laws we can fully fulfill. Technoscientists should strive to promise to do their best, realize when they fall short, and then try to make up for their shortcomings or failures when they occur. There is something naïve about such an expectation, but this alone is no reason to abandon the hope that technoscientists will indeed be trained more broadly to consider social issues beyond the technical aspects of their expertise.

If it's too optimistic to imagine technoscientific practice in ideal terms, let's still insist that the scientific enterprise will remain a scientific community, however flawed. At least as a community, it can offer institutional procedures to self-examine how well or poorly it can fulfill its own promises and ideals. When the public nourishes its technoscientific community – with prestige and monetary rewards – it should not only hope for a certain code of behavior, but also ensure that democratic principles are part of the conversation. Technoscience as vocation is a human aspiration, a worthy one at that. Universalism, communalism, disinterestedness, and organized skepticism, as well as originality or novelty and personal integrity are ideals worthy of implementation. And ethical injunctions as guidelines are much easier to uphold than we imagine. Together, all of these interrelated notions and practices should inform the integrity of a human endeavor we cannot afford to hand over to the gods of the polis or of mammon.

DOI: 10.1057/9781137519429.0008

References

Joseph Agassi (1981), *Science and Society: Studies in the Sociology of Science*. Dordrecht, Boston, and London: D. Reidel Publishing Company.

Joseph Agassi (2003), "Science and Its Public Relations," *Science and Culture*. Dordrecht, Boston, and London: Kluwer Academic Publishers, Chapter 2.6, pp. 152–163.

Ivan Amato, "Rustum, Roy (1992), "PR Is a Better System than Peer Review." *Science*. 258: 736.

Marcia Angell (1997/1996), *Science on Trial: The Clash of Medical Evidence and the Law in the Breast Implant Case*. New York: W. W. Norton & Company.

Marcia Angell (2004), *The Truth about Drug Companies: How They Deceive Us and What to Do about It*. New York: Random House.

John Ashton and Ron Laura (1998), *The Perils of Progress: The Health and Environment Hazards of Modern Technology, and What You Can Do about Them*. London and New York: Zed Books Ltd.

Charles Babbage (2013/1830), *Reflections on the Decline of Science in England, and on Some of Its Causes*. San Bernardino, CA: Hard Press.

Francis Bacon (1985/1620), *The New Organon and Related Writings*. New York: Macmillan Publishing Company.

David M. Berry (2008), *Copy, Rip, Burn: The Politics of Copyleft and Open Source*. London: Pluto Press.

Elizabeth Popp Berman (2012), *Creating the Market University: How Academic Science Became and Economic Engine*. Princeton and Oxford: Princeton University Press.

Elizabeth Popp Berman (2014), "Not Just Neoliberalism: Economization in US Science and Technology Policy." *Science, Technology, & Human Values* 39 (3): 397–431.

Vadim J. Birstein (2001), *The Perversion of Knowledge: The True Story of Soviet Science.* Boulder: Westview Press.

Eula Biss (2014), "The Illusion of 'Natural.'" *The Atlantic.* http://www.theatlantic.com/health/archive/2014/09/the-illusion-of-natural/380836/. Accessed December 19, 2014.

Michael Bowker (2003), *Fatal Deception: The Untold Story of Asbestos.* New York: Rodale.

Chris Bowlby (2011), "Fritz Haber: Jewish Chemist Whose Work Led to Zyklon B." *BBC Radio 4* (April 11, 2011).

Peter R. Breggin (1991), *Toxic Psychiatry: Why Therapy, Empathy and Love Must Replace the Drugs, Electroshock, and Biochemical Theories of the "New Psychiatry."* New York: St Martin's Press.

William J. Broad (1983), "Physicists Compete for the Biggest Project of All." *The New York Times* (September 20, 1983).

William Broad and Nicholas Wade (1982), *Betrayers of the Truth: Fraud and Deceit in the Halls of Science.* New York: Simon & Schuster, Inc.

Shannon Brownlee (2007), *Overtreated: Why Too Much Medicine Is Making Us Sicker and Poorer.* New York: Bloomsbury.

Rebecca Culshaw (2007), *Science Sold Out: Does HIV Really Cause AIDS?* Berkeley, CA: North Atlantic Books.

Robert Dahl (1985), *A Preface to Economic Democracy.* Berkley and Los Angeles: University of California Press.

Catherine D. DeAngelis (2006), "The Influence of Money on Medical Science." *Journal of the American Medical Association* 296(8): 996–998 (August 23/30, 2006).

Steve Fuller (2004), *Kuhn vs. Popper: The Struggle for the Soul of Science.* New York: Columbia University Press.

Steve Fuller (2007), *Science vs. Religion? Intelligent Design and the Problem of Evolution.* Cambridge, UK: Polity Press.

Damian Garde (2014), "Billionaire Soon-Shiong Dives into Immuno-Oncology with $20M Sorrento Venture." *Fierce Biotech.* http://www.fiercebiotech.com/story/billionaire-soon-shiong-dives-immuno-oncology-20m-sorrento-venture/2014-12-15. Accessed December 15, 2014.

Gerd Gigerenzer (2002), *Calculated Risks: How to Know When Numbers Deceive You.* New York: Simon & Schuster.

DOI: 10.1057/9781137519429.0009

Stephen Jay Gould (1977), *Ever Since Darwin: Reflections in Natural History.* New York: Norton.

John Grant (2007), *Corrupted Science.* Surrey, UK: AAPPL Artists' and Photographers' Press Ltd.

Daniel S. Greenberg (2001), *Science, Money, and Politics: Political Triumph and Ethical Erosion.* Chicago and London: The University of Chicago Press.

Daniel S. Greenberg (2007), *Science for Sale: The Perils, Rewards, and Delusions of Campus Capitalism.* Chicago and London: The University of Chicago Press.

Catherine Ryan Hyde (1999), *Pay It Forward.* New York: Simon & Schuster.

Henrik Ibsen (1970/1882), *An Enemy of the People.* New York: New American Library.

Alan Irwin (1995), *Citizen Science: A Study of People, Expertise and Sustainable Development.* London and New York: Routledge.

David Joravsky (1970), *The Lysenko Affair.* Chicago & London: The University of Chicago Press.

Horace Freeland Judson (2004), *The Great Betrayal: Fraud in Science.* New York: Harcourt, Inc.

Daniel Kahneman (2012), *Thinking, Fast and Slow.* New York: Farrar, Straus, and Giroux.

Clark Kerr (1995/1963), *The Uses of the University.* Cambridge, MA: Harvard University Press

Sheldon Krimsky (2003), *Science in the Private Interest: Has the Lure of Profits Corrupted Biomedical Research?* London and Boulder: Rowman & Littlefield Publishers, Inc.

Thomas S. Kuhn (1970/1962), *The Structure of Scientific Revolutions.* Chicago: The University of Chicago Press.

Bruno Latour and Steve Woolgar (1979), *Laboratory Life: The Construction of Scientific Facts.* Princeton, NJ: Princeton University Press.

Mario Livio (2013), *Brilliant Blunders: From Darwin to Einstein—Colossal Mistakes by Great Scientists that Changed Our Understanding of Life and the Universe.* New York: Simon & Schuster.

Joseph P. Martino (1992), *Science Funding: Politics and Porkbarrel.* New Brunswick and London: Transaction Publishers.

Mariana Mazzucato (2014), *The Entrepreneurial State: Debunking Public vs. Private Sector Myths.* London and New York: Anthem Press.

Barry Meier (2013), "Johnson & Johnson in Deal to Settle Hip Implant Lawsuits." *The New York Times* (November 19, 2013).

DOI: 10.1057/9781137519429.0009

Robert Merton (1965). *On the Shoulders of Giants*. Chicago and London: University of Chicago Press.

Robert K. Merton (1973/1942), "The Normative Structure of Science," *The Sociology of Science: Theoretical and Empirical Investigations*. Chicago and London: The University of Chicago Press, pp. 267–278.

Stanley Milgram (1974), *Obedience to Authority: An Experimental View*. New York: Harper Collins.

Philip Mirowski (2011), *Science-Mart: Privatizing American Science*. Cambridge and London: Harvard University Press.

Gerald C. Nelson (2001), *Genetically Modified Organisms in Agriculture: Economics and Politics*. Academic Press.

Marion Nestle (2002), *Food Politics: How the Food Industry Influences Nutrition and Health*. Berkeley: University of California Press.

Paul A. Offit and Sarah Erush (2013), "Skip the Supplements." *The New York Times* (December 14, 2013).

Eric Palmer and Carly Helfand (2014), "The Top 10 Pharma Companies by 2013 Revenue." *Fierce Pharma* (March 4) http://www.fiercepharma.com/special-reports/top-10-pharma-companies-2013-revenue. Accessed May 10, 2014.

Stuart Parkinson and Chris Langley (2009), "Stop the Sell-Out." *New Scientist* 204(2733): 32–33.

Thomas Piketty (2014), *Capital in the Twenty-First Century*. Translated by Arthur Goldhammer. Cambridge and London: Harvard University Press.

Michael Polanyi (1966), *The Tacit Dimension*. New York: Doubleday & Company.

Andrew Pollack (2013), "Doctors Say Heart Drug Raised Risk of an Attack." *The New York Times* (November 19, 2013).

Karl R. Popper (1962), *Conjectures and Refutations: The Growth of Scientific Knowledge*. New York: Basic Books.

Neil Postman (1992), *Technopoly: The Surrender of Culture to Technology*. New York: Vintage Books.

Eyal Press and Jennifer Washburn (2000), "The Kept University." *The Atlantic* (March 1, 2000).

Robert N. Proctor (1995), *Cancer Wars: How Politics Shapes What We Know and Don't Know about Cancer*. New York: Basic Books.

Sheldon Rampton and John Stauber (2001), *Trust Us, We're Experts! How Industry Manipulates Science and Gambles with Your Future*. New York: Penguin Putnam, Inc.

DOI: 10.1057/9781137519429.0009

Kal Raustiala and Christopher Sprigman (2012), *The Knockoff Economy: How Imitation Sparks Innovation*. Oxford and New York: Oxford University Press.

Jeremy Rifkin (2014). *The Zero Marginal Cost Society: The Internet of Things, the Collaborative Commons, and the Eclipse of Capitalism*. Hampshire, UK: Palgrave Macmillan.

Marie-Monique Robin (2010), *The World According to Monsanto: Pollution, Corruption, and the Control of Our Food Supply*. New York: The New Press.

Raphael Sassower (2009), *Postcapitalism: Moving beyond Ideology in Economic Crises*. Boulder: Paradigm Publishers.

Raphael Sassower (2013a), "Technoscience and Society." *Encyclopedia of the Philosophy and the Social Sciences* (edited by Byron Kaldis). New York: Sage Publications, Inc.

Raphael Sassower (2013b), *Digital Exposure: Postmodern Postcapitalism*. Hampshire, UK: Palgrave Macmillan.

Raphael Sassower (2014), *The Price of Public Intellectuals*. Hampshire, UK: Palgrave Macmillan.

James D. Savage (1999), *Funding Science in America: Congress, Universities, and the Policies of the Academic Pork Barrel*. Cambridge: Cambridge University Press.

Howard K. Schachman (1993), "What Is Misconduct in Science?" *Science* 261(July 9, 1993): 148–153.

Eric Schlosser (2001), *Fast Food Nation: The Dark Side of the All-American Meal*. New York: Houghton Mifflin Company.

Alan Schwartz (2013), "The Selling of Attention Deficit Disorder." *The New York Times* (December 14, 2013).

Steven Shapin (1994), *A Social History of Truth: Civility and Science in Seventeenth-Century England*. Chicago and London: The University of Chicago Press.

Katherine Sharpe (2012), *Coming of Age on Zoloft: How Antidepressants Cheered Us Up, Let Us Down, and Changed Who We Are*. New York: Harper Perennial.

Upton Sinclair, Jr. (1905), *The Jungle*. New York: Harper & Brothers Publishers.

T. D. Singh and Ravi V. Gomatam (1987), "Is Science Selling Out?," Interview with Maurice H. Wilkins, *Synthesis of Science and Religion: Critical Essays and Dialogues*. San Francisco: Bhaktivedanta Institute.

DOI: 10.1057/9781137519429.0009

http://www.vedicsciences.net/articles/science-selling-out.html.
Accessed May 2, 2014.

Rebecca Skloot (2011/2010), *Immortal Life of Henrietta Lacks*. New York: Random House.

Sheila Slaughter and Larry L. Leslie (1997), *Academic Capitalism: Politics, Policies, and the Entrepreneurial University*. Baltimore and London: The Johns Hopkins University Press.

Robert Shiller (2005), *Irrational Exuberance* (2nd ed.). Princeton NJ: Princeton University Press.

Adam Smith (1976/1759), *The Theory of Moral Sentiments*. Indianapolis: Liberty Classics.

Adam Smith (1937/1776), *An Inquiry into the Nature and Causes of the Wealth of Nations*. Edited by E. Cannan. New York: The Modern Library.

Jeffrey M. Smith (2007), *Genetic Roulette: The Documented Health Risks of Genetically Engineered Foods*. Yes! Books.

Andrew Stark (2000), *Conflict of Interest in American Public Life*. Cambridge and London: Harvard University Press.

Clifford Stott (2009), "How Violent Are You?" *Horizon* Series 45, Episode 18, BBC.

Thomas S. Szasz (1974/1960), *The Myth of Mental Illness: Foundations of a Theory of Personal Conduct*. New York: Harper & Row.

Nick Paul Taylor (2014), "Billionaire Doc Teams with Oxford Uni to Build 'NASA of Biomedicine.'" *Fierce Biotech*. http://www.fiercebiotechit.com/story/billionaire-doc-teams-oxford-uni-build-nasa-biomedicine/2014-10-27. Accessed December 15, 2014.

Katie Thomas (2013a), "J & J to Pay $2.2 Billion in Risperdal Settlement." *The New York Times* (November 4, 2013).

Katie Thomas (2013b), "Glaxo Says It Will Stop Paying Doctors to Promote Drugs." *The New York Times* (December 16, 2013).

Charles Thorpe and Jane Gregory (2010), "Producing the Post-Fordist Public: The Political Economy of Public Engagement with Science." *Science as Culture* 19(3): 273–301, September 2010.

Eric Topol (2012), *The Creative Destruction of Medicine: How the Digital Revolution Will Create Better Health Care*. New York: Basic Books.

Karl Weber (2009), *Food, Inc.: How Industrial Food Is Making Us Sicker, Fatter and Poorer—And What You Can Do About It*. New York: Public Affairs.

Max Weber (1946/1919), "Science as a Vocation," in *From Max Weber: Essay in Sociology*, edited by H. H. Gerth and C. Wright Mills. New York: Oxford University Press, pp. 129–156.

DOI: 10.1057/9781137519429.0009

Alvin M. Weinberg (1961), "Impact of Large-Scale Science on the United States." *Science* 134 (3473): 161–164 (July 21, 1961).

John Ziman (2000), *Real Science: What It Is, and What It Means.* Cambridge: Cambridge University Press.

"Trouble at the Lab." *The Economist* (October 19, 2013). http://www. economist.com/news/briefing/21588057-scientists-think-science-self-correcting-alarming-degree-it-not-trouble. Accessed October 22, 2013.

DOI: 10.1057/9781137519429.0009

Index

DOI: 10.1057/9781137519429.0010